HERE'S WHAT REVIEWERS ARE SAYING ABOUT

HEALTHY WATER FOR A LONGER LIFE

"Excellent book . . . I think you have done a good job in trying to put this all together in one single source."

Jeffrey Bland, Ph.D., Director
Laboratory for Nutritional Supplement Analysis
Linus Pauling Institute of Science & Medicine

"Its the best I've read on this subject and I intend to recommend it."

W. Marshall Ringsdorf, Jr., D.M.D.
Author: The Vitamin C Connection

". . . Your book appears to probably be the most concise, authoritative summary work yet on the subject."

Robert T. Williams, President
General Ecology, Inc.

"I read your excellent book, Healthy Water For A Longer Life with great pleasure! I highly recommend it's thoughtful, informative contents to any one concerned reader."

Gene Rosov, President
WaterTest Corporation

"The best documented water book yet for the average citizen to read."

Warren Clough, Analytical Chemist
Ozark Water Service & Analytical Laboratory

"An easy-to-read and very up-to-date compendium of much of the present literature, scientific and otherwise, regarding drinking water."

Joseph M. Price, M.D.
Author: Coronaries/Cholesterol/Chlorine

"This is truly a fabulous, as well as informative work. Its 144 pages are crammed full of very valuable data on water."

Multipure Drinking Water Systems

"This is a researched and well documented informational book. Long overdue, it tells the story of how bad our drinking water is to our health. Virtually all other books on drinking water have been based on opinions and Dr. Fox has done an excellent job of 'telling it like it is' in regard to our health."

The Clinical Nutritionists Newsletter

"Here, at last, is a comprehensive book that cuts through the hearsay in answering intelligently the water question . . . This is a 'must read' for anyone who cares what he is putting into his body."

June Peterson, Health World
World Media Syndicate

Healthy Water for a Longer Life

By

Martin Fox, Ph.D.

Preface By

Robert D. Milne, M.D.

and

Andrew C. Stenhouse, M.D.

DUNAWAY FOUNDATION
Amarillo Texas

Second Edition, August 1986

ISBN# 0-9617432-0-4

Cover design by Holly Landgarten Fox

Printed in the United States of America

Healthy Water Research
29 Mendum Avenue
Portsmouth, NH 03801
433-3360

All inquires should be addressed to:

Healthy Wat[illegible]h
Route 6, [illegible]
Austin, Te[illegible]

For additional copies, send $7.95 plus $2.00 shipping.

Audio Cassettes:

- "Healthy Water For A Longer Life: A Summary" Overview and summary of the main ideas presented in this book. 45 minutes. $7.95 plus $2.00 shipping.

- "Chlorination: A Link Between Heart Disease & Cancer" 10 minutes. $5.95 plus $2.00 shipping.

Quantity discounts available.

• CONTENTS

• CONTENTS

TABLES

• PREFACE

From a scientific point of view, our society is overwhelmed by the remarkable explosion in knowledge and investigations which the advances in biology and physics have provided medicine.

We are sometimes dismayed that even with all its sophisticated technology, medical care many times falls short in helping man's suffering and anguish. Many times we miss the important simple truths. One of those must be water - the forgotten nutrient.

Dr. Fox has thoroughly researched the available studies and they have admirably been presented. Some conclusions on water are difficult when important studies still await clarification of important issues.

However, one thing is clear: pure water is a necessity and it will become a major health issue for the twenty-first century. We thank Dr. Fox for reminding us of the basics.

Robert D. Milne, M.D.
Andrew C. Stenhouse, M.D.

• INTRODUCTION

Each of us has at one time or another approached a subject with preconceived notions. Yet, once we study that topic in depth, we are amazed to find that our earlier held beliefs were mistaken. To some degree, this pattern describes my own examination of drinking water.

The most surprising discovery for me was that virtually all the available books on drinking water are based on opinions rarely documented with research. And most of the research on water is fragmented and does not integrate related findings. For instance, the literature studying drinking water and its relationship to heart disease ignores the studies on drinking water and cancer - and vice versa.

In this book, I have attempted to pull together the various aspects of research on water. I have tried to look at the entire body of literature on drinking water in several health areas and document the emerging connections. I wanted to develop a complete picture that would address all of the major concerns about this precious substance so absolutely vital to the life and good health of each of us.

I believe there is strong evidence to support the contention that drinking the correct type of water is a missing and vital link in our overall health plan. Specifically, drinking water containing the proper amounts of important minerals can significantly lower our risk of heart disease and cancer! By drinking healthy water each of us will enjoy longer life.

I want to thank several people and institutions for their support in the evolution of this book.

Thanks to Rae and Gerald Landgarten for their financial help when it was most needed; Robert D. Milne, M.D., and Andrew C. Stenhouse, M.D. for their constant encouragement and support; The Dunaway Foundation for their financial assistance; Kurt Donsbach, Ph.D. and Donsbach University; Neva Jensen, Herbalist; Betty Lee Morales, Ph.D., Nutritionist; The Reference Librarians of the Chula Vista Public Library, Chula Vista, California; and Don Snow for his useful suggestions and steady friendship.

And special thanks to my wife, Holly Landgarten Fox, for her editing, typing, book design, love and patience.

1 • THE WATER STORY AND HEART DISEASE

Research into the effect of water on heart disease mainly began in 1960 with Schroeder. In his paper, "Relation Between Mortality from Cardiovascular Disease and Treated Water Supplies," the water in 163 largest cities in the United States was analyzed. The analysis was for twenty-one constituents and correlated to cardiovascular disease (CVD).

Schroeder found softer water was associated with higher death rates. "Some factor either present in hard water or missing or entering soft water appears to affect death rates from degenerative cardiovascular disease."(1)

Once this study was published, the search was on to see if Schroeder's findings could be duplicated. And, if confirmed, to understand the mechanism of how water could play such an important role in heart disease. In the twenty or more years that have passed, a vast amount of research has occured.

Before we review and discuss the research, let's understand what is generally meant by hard water.

> Water hardness is most often expressed as the amount of calcium carbonate or its equivalent, expressed as mg of CaCo3 per liter of water . . . In most systems, calcium and magnesium compounds in varying proportions make up the bulk of hardness. The four categories of hardness in common use are: 0-60 ppm, soft; 61-120 ppm, moderately hard; 121-180 ppm, hard; 181 ppm or more, very hard; where hardness is expressed as CaCo3. (2)

Neri searched for a water factor by studying large population groups (epidemiology). He found three important facts which appear to be well established:

> (1) . . . there are geographical differences in the distribution of mortality, both in 'all causes' and cardiovascular disease.
> (2) . . . there is a definite association between water quality and mortality from various disease categories - in particular cardiovascular.
> (3) . . . increase in the use of manufactured foods . . . refining and processing of food leads to loss of important amounts of nutrients, and particularly trace metals . . . Parallel with this increased usage of 'nutrient deprived' foods is the present epidemic of premature mortality from cardiovascular disease. (3)

Notice that Neri talks about an "association between water quality and mortality" and not a cause and effect relationship. But, the repeated correlations of water quality with mortality is highly significant.

In 1979, Comstock reviewed and commented on twenty years of water research in the American Journal of Epidemiology. He sites close to fifty studies and evaluates each study looking at the epidemological criteria: consistency, strength of association, dose-response effect, specificity, experimental changes, confounding factors and biological plausibility.

After reviewing these studies,

> . . . there can be little doubt that the associations of water hardness with cardiovascular mortality are not spurious. Too many studies have reported statistically significant correlations

> to make chance or sampling errors a likely explanation. The complete independence of virtually all the observations of water hardness and the measures of CVD effectively rules out observer bias as a cause of the relationship. (4)

> . . . the most likely effect of a water factor is mediated either through a deficiency of an essential element or an excess of a toxic one.(5)

After twenty years, we are left with Schroeder's original insight: hard water is better than soft water. Comstock suggests it may be due to a deficiency of an essential element or an excess of a toxic element. Or it could be a combination of both!

Sharrett looked at geographical and human tissue studies. Like Comstock, Sharrett's analysis finds hard water associated with less CVD and soft water associated with a higher rate of CVD. Even though studies on specific elements have shown inconsistent findings.

> The available geographic studies relating water quality to mortality rates have not provided consistent evidence regarding the effects of elements found in drinking water on the cardiovascular system . . . the association of low mortality rates in several countries with hard water might reflect the importance of calcium and magnesium, which are the major components of hardness, but it give no clue to the possible importance of chromium, copper, zinc, cadmium or lead because the relationship of these constituents to hardness varies from place to place.(6)

Sharrett analyzed geographic studies and found data relating specific elements in water to death rates was not consistent. Yet, hard water (high in calcium and magnesium) seems to result in lower mortality.

The connection between hardness and trace elements is unclear.

> None of the trace elements studied, however, appears to be associated with hardness closely enough to explain the hardness-mortality associations, though there are insufficient data to conclude that there is no association. (7)

The trace elements studied were chromium, copper, zinc, cadmium and lead.

Sharrett tried to locate the specific elements that may be responsible for the beneficial effects of hard water. He looked at the relationship of elements in drinking water with the elements in human tissues. The elements he studied in human tissues were magnesium, chromium, copper, zinc, cadmium and lead.

After reviewing the studies, Sharrett concludes:

> There is little evidence bearing on the question of a substantial contribution of drinking water to human tissue levels of cadmium, chromium, or zinc. Copper and magnesium levels of tissues may be related to drinking water . . . Lead levels in blood and possibly other tissues, are almost certainly affected by lead levels in drinking water in areas where these levels are particularly elevated . . . (8)

He finds that copper, magnesium and lead may be important in understanding the role of water to health. However, in most cases, there were not enough well designed studies to test the importance of trace elements. Also, the synergistic effect of hard water coupled with trace elements is not carefully considered. In the chapter, The Water Story and Animal Experiments, we will see a definite synergistic effect between water hardness and other minerals.

Sharrett overlooked the studies on tissue levels of calcium. Because hard water usually contains a significant amount of magnesium, as well as calcium, the following studies discuss both. In "Cardiovascular Disease and Minerals in Drinking Water," Crawford summarizes three studies.

> (1) The calcium and magnesium content of the arteries in young men was found to be significantly lower in soft water areas.
> (2) Mean values for serum calcium and magnesium were lower in men living in a town using soft water.
> (3) Mean values for total serum calcium in a farming village with high cerebrovascular death-rates were lower than in a fishing village with low rates . . . (9)

These studies show definite differences in tissue levels of calcium and magnesium between soft water and hard water areas. They lend physiological support to the epidemiological studies showing a decrease in heart disease mortality in hard water areas.

A different approach to understand the value of minerals in drinking water comes from Sauer. Rather than try to narrow the water factor to a specific element, Sauer looked at water from a broader perspective. In his "Relationship of Water to the Risk of Dying," he studied the relationship of total dissolved solids (TDS) to heart disease and to other chronic diseases.

Chronic diseases account for about nine-tenths of all deaths in adults in the United States.

> . . . diseases of the heart and blood vessels, various forms of cancer or malignant neoplasms, diabetes, cirrhosis of the liver, chronic respiratory diseases and others. (10)

Sauer used twenty-three measures of characteristics of the drinking water in ninety-two metropolitan state economic areas.

> Total dissolved solids in the drinking water consistently showed the highest or one of the highest correlations with death rates as shown in Table 1. Those at the .01 level, by a conventional standard, may be considered statistically significant . . . The associations are usually negative - that is, the higher the content of dissolved solids or of specific substances (Table 2, 3 & 4), the lower the death rate for the cardiovascular-renal (CVR) diseases as a group, for coronary heart disease specifically, and for all forms of cancer as a group. Hypertensive heart disease, stroke, and other cardiovascular-renal diseases tend to have negative correlations also, but at levels too low to be considered significant. (11)

TABLE 1

Correlation, total dissolved solids in drinking water and death rates, whites age 35-74

	Male	Female
All causes	-.32	-.30
CVR diseases	-.44	-.39
Coronary	-.46	-.44

TABLE 1 (Continued)

	Male	Female
Cancer	-.42	-.29
Accidents	+.32	+.26
Chr. respiratory	+.34	+.29

TABLE 2

Correlation of silica (Si02) and death rates, whites age 35-74

	Male	Female
CVR diseases	-.28	-.26
Coronary	-.37	-.34
Stroke		-.35
Cancer	-.43	-.42
Accidents, viol.	+.50	
Chr. respiratory	+.35	

TABLE 3

Correlations with death rates, whitesage 35-74

MALE Rates	Hardness	Sodium
All causes	-.33	
CVR diseases	-.35	-.36
Coronary	-.35	-.39
Cancer	-.37	-.31
Accidents, viol.		+.30
FEMALE Rates		
All causes		-.26
CVR diseases	-.28	-.34
Coronary	-.31	-.39
Stroke		-.29
Cancer		-.33

TABLE 4

Correlation, trace elements and death rates, whites age 35-74

MALE Rates	Strontium	Lithium	Vanadium
CVR diseases	-.34	-.32	-.36
Coronary	-.39	-.39	-.40
Cancer	-.33	-.34	-.34
Chr. respiratory	+.39	+.35	
FEMALE Rates			
CVR disease	-.32		
Coronary	-.39	-.33	-.30
Stroke			-.41
Chr. respiratory	+.36	+.33	

Science tries to separate, analyze, reduce things to a single cause and see if this localized item can create the desired effect. Water hardness is a measurement of CaCo3 or calcium and magnesium. Hardness is a broader concept than just a specific individual mineral, e.g. sodium, zinc, copper and so forth. However, the concept of total dissolved solids is an even more inclusive concept than hardness.

Total dissolved solids (TDS) includes all the various elements in water. Very few studies have looked at TDS. Most studies prefer instead to select either calcium and magnesium as a measure of hardness, or a specific trace element to study. Perhaps this is one of the reasons why there are inconsistencies in the water story. Looking at the whole and not simply the parts, we may get a truer picture of the beneficial nature of water.

Sauer's studies are unique. He looked at the total dissolved solids and so, succeeded in expanding the scope of the way we look at water. His studies clearly show the value of a higher content of total dissolved solids in water. The higher the total dissolved solids, the lower the death rate from cardiovascular diseases and cancer.

The more we try to isolate and study a single water factor, the more inconsistent the results can be. The most consistent results have come from studying water hardness as opposed to studying individual elements.

We can expect even greater consistency to result from looking at total dissolved solids (TDS) with the understanding that as TDS increases, chronic diseases come down. This is what Sauer's study shows.

The synergistic effect of the TDS on the body is probably two complex to localize. Perhaps this is why the European mineral waters, which have been used for centuries for their health properties, have been so popular and beneficial. If we look at these mineral waters and analyze them to see which specific minerals are predominant, we find that these waters are very different and unique. The only overall pattern seems to be that they are high in total dissolved solids. In Chapter 3, The Water Story and Cancer, the importance of TDS will take on even greater impact.

Sometimes, the best experiments are those that nature has been conducting for years, yielding information that may not be obtainable by our limited experiments. In this regard, there are two studies which deserve to be highlighted. They show two pairs of cities, each with hard water, with one of the cities altering the water to create soft water. What was the effect of softening their water? A higher rate of heart disease!

The first example comes from England and the neighboring towns of Scunthrope and Grimsby. Both towns drank the same water for years. Scunthrope softened its water from 488 ppm of total hardness to 100 ppm. Grimsby did not change its water. The results:

There appears to have been a striking increase in CVD in the town that softened its water, whereas, if anything, in the other town either nothing happened or there was a slight increase. (12)

The second example comes from Italy and the two rural areas of Crevalcore and Montegiorgio. The following summary is from a study reported in 1980 in the American Heart Journal. (13)

In Crevalcore, the water has been supplied since 1915 by an aqueduct with a water hardness of 215 mg/L CaCo3. Montegiorgio used well and spring waters containing 442 mg/L of CaCo3. In 1960, construction started on an aqueduct in Montegiorgio which supplied the people with a water hardness of 131 mg/L CaCo3. The change of water supply from wells to aqueduct transformed Montegiorgio from an area supplied with very hard water to one with a softer kind of water, similar to the water supplied for a long time in Crevalcore.

At the beginning of the survey, there was a higher prevalence of heart disease and hypertension and a higher incidence of mortality rate for heart disease in Crevalcore. Since the aqueduct in Montegiorgio was constructed, the heart disease mortality rate during the last five years is definitely higher than in Crevalcore.

From both of these studies, we observe that the harder the water, the less the heart disease. What is surprising is that in both studies, the changed water supply still provided relatively hard water: in Scunthrope, the water is 100 ppm of total hardness and Montegiorgio's water is 131 mg/L of CaCo3. However, in both cases, the change was over a 20% decrease in water hardness.

In England, the British Regional Heart Study has reported their initial findings on cardiovascular mortality and the role of water quality. The study looked at cardiovascular mortality in 253 English towns during 1969 to 1973. The results were significant: A clear, negative association between water hardness and cardiovascular mortality.

> . . . A significant negative association remains between water hardness and cardiovascular mortality. The relationship between water hardness and cardiovascular mortality is nonlinear, and there appears to be a 10-15% excess of cardiovascular deaths in areas with very soft water compared with areas of medium hardness. (14)

In soft water areas (25 mg/L CaCo3) compared to hard water areas (170 mg/L CaCo3), there was 10% to 15% more cardiovascular deaths. There was no evidence that water harder than 170 mg/L would result in lower CVD mortality. (15)

In light of all these studies on softening water, it is interesting to note that the towns which softened their water - Scunthrope and Montegiorgio - had a water hardness of 100 and 131 mg/L respectively!

At this level of hardness, the towns experienced an increase in heart disease. Perhaps this effect would disappear if the water had a hardness of 170 mg/L as the British Study indicates!

In the United States, a report similar to the British Regional Heart Study was undertaken by Greathouse and Osborne. They surveyed 4200 adults aged 25 to 74 from thirty-five geographic areas. The random population selection represented an analysis of the entire civilian population.

> Each participant was interviewed and given a thorough physical examination. A tap water

> grab sample ws collected from each participants' residence and analyzed for 80 inorganic chemical constituents . . . Hardness and calcium appear to follow the normal trend of negative associations with the mortality rates for most groups of cardiovascular diseases . . . (16)

The National Academy of Sciences a few years ago stated "optimal conditioning of drinking water could reduce the amount of cardiovascular disease mortality rate by as much as 15% in the United States." (17)

These studies, coupled with the epidemiological studies discussed earlier, strongly support the belief that there is a strong and important relationship between water hardness, total dissolved solids (TDS) and heart disease. Hard water and water high in TDS is more beneficial and protective to your heart than soft water and water low in TDS. In addition, it seems the water hardness should be around 170 mg/L of CaCo3 to be most effective. Guidelines for total dissolved solids will be discussed in Chapter 3, <u>The Water Story and Cancer</u>.

2 • SODIUM, HYPERTENSION AND DRINKING WATER

> Hypertension is the most common of the chronic diseases in the United States, afflicting some 24 million people . . . It is a major problem in all the developed nations. (1)

Much research has been done on the relationship of sodium and blood pressure. An excellent, thorough review of the diverse findings can be found in Blackburn and Prineas' article, "Diet and Hypertension: Anthropology, Epidemology, and Public Health Implications." A very brief summary follows.

The sodium chloride (salt) intake for Western cultures is 8 to 15 grams daily. This includes inhabitants of continental Europe, Scandinavia, the British Isles, Australia, New Zealand and North America.

After evaluating the research world-wide, the authors recommend a sodium intake of 2 to 5 grams (g) daily.

> . . . an 'ideal' population intake of salt would be a median value around 3 g daily with a range of 2 to 5 g. This distribution is compatible with normal growth and development, with physiological function and with the absence of adult hypertension. (2)

A reduced salt intake can lower high blood pressure and there is evidence that a low salt diet can prevent high blood pressure in humans. Many factors are involved in high blood pressure besides sodium. Diets high in potassium and calcium, diets rich in vegetables with less meat and fat consumption have ben shown to be effective in reducing and/or preventing high blood pressure. (3)

Even though it is estimated by the National Academy of Sciences that 90 percent of our sodium intake is from food and 10 percent from drinking water, several studies have started to look at the relationship of water sodium and blood pressure.

One of the first reports to study this relationship was conducted by Tuthill and Calabrese. They surveyed two groups of Massachusetts high school students in matched communities. One group had drinking water with 107 mg/L of sodium; the other group had 8 mg/L of sodium in their drinking water.

> High school sophomores residing in a community with elevated levels of sodium in the drinking water (107 mg/L) exhibited a marked upward shift in blood pressure distribution patterns for systolic and diastolic pressure, when compared with students in an appropriately matched community (8 mg/L). Females exhibited a blood pressure distribution pattern characteristic of persons 10 years older, while for males the upshift was similar to that of a group approximately 2 years older. (4)

A follow-up study in the same two communities was conducted on third grade students. This study considered dietary intake and urinary excretion of sodium. The results were similar to the previous study. Increased sodium levels in drinking water resulted in increased blood pressure levels.

Stimulated by these findings, several researchers conducted studies similar to Tuthill's but usually having different results. However, before we leave Tuthill's report, we need to be aware of potential difficulties in his study design and procedures. One critic points out that the probability of chance, observational bias and statistical methodology could have 'produced' the observed changes in blood pressure instead of the water sodium!(5) In any case, most studies do not find significant evidence that water sodium levels affect blood pressure. In light of the critical remarks concerning Tuthill's design, and the studies to be shortly discussed, we would be wise to interpret the Massachusett's findings with caution.

In Australia, Armstrong and associates studied 326 boys and 309 girls, aged 12 to 14, in six different towns. The range of sodium in the drinking water was nearly twice the range in the studies by Tuthill. The sodium range was 8 to 107 mg/L in Tuthill's study and in the Armstrong study, the range was .13 to 9.69 mmol/L. (Armstrong used mmol/L - millimoles per liter - instead of mg/L - milligrams per liter - in quantifying the amount of sodium in the water. 100 mg equals 4.3 mmol of sodium. Therefore, the .13 to 9.69 mmol/L equals 3 to 225 mg/L.)

> This study has shown no evidence of a relationship between blood pressure in children 12 to 14 years of age with water sodium levels ranging from 0.13 to 9.69 mmol/L. There was significant differences in mean blood pressure between children in the towns studied, but they showed no consistent relationship with water sodium levels and did not appear to be explained by differences between the towns in distributions of variables such as total sodium and potassium intake; dietary meat, fat and energy adiposity; and physical activity. (6)

However, another study was conducted in Holland with somewhat similar results as Tuthill's Massachusetts study. Sodium levels influenced the blood pressure in 348 children, 7 to 11 years old. Three towns with different water sodium levels were studied. The long-term low town had low water sodium levels and the long-term high town had high water sodium levels. In both of these towns, water sodium levels had not changed for the past fifteen years. The third town, refered to as the short-term high area in the study, had its water sodium level changed one year before the study from 1 to 7 mmol/L (23 to 163 mg/L).

> The mean values of systolic and diastolic blood pressure were higher in the high sodium areas . . . The findings . . . support the hypothesis that sodium intake influences blood pressure. The association seems to be of a relatively short-term nature, as no difference in blood pressure levels were found between long-term and short-term high areas. (7)

Higher water sodium levels resulted in higher blood pressures. However, no significant differences were found in mean blood pressure between the children living in the long-term and short-term high towns. This implies the increase in blood pressure occurs quickly but additional exposure does not seem to result in further increases in blood pressure.

A similar study was conducted in the two communities of LaGrange and Westchester in the Chicago metropolitan area. Both of these towns have similar socio-economic and demographic characteristics. The main difference was in the water sodium levels. The range was 405 mg/L in LaGrange and 4 mg/L in Westchester. This range was substantially higher than the towns in Massachusetts with 107 mg/L and 8 mg/L of sodium.

The blood pressure of the high school junior and seniors from the two Chicago towns were compared. The findings do not support the results of Tuthill's study.

> Results of the survey indicated that male and female systolic blood pressures in the high sodium community were not significantly higher than those in the low sodium community. Surprisingly, the observed systolic blood pressure of the males in the low sodium community were higher than those in the high sodium community. These findings did not corroborate the results of the Massachusetts study. (8)

Although the systolic blood pressure was not effected by the amount of water sodium, the diastolic blood pressure was.

> However, the male and female diastolic blood pressure were significantly higher . . . in the high sodium community. The increases in diastolic blood pressures . . . were not as large as those observed in the Massachusetts study . . . (9)

The effects of drinking water on blood pressure was studied in 295 persons in rural Michigan. Again, there were no significant relationships between water sodium and blood pressure levels or hypertension.

> Sodium in drinking water, dietary sodium intake, blood pressure, sodium excretion, height and weight were measured. No significant relationships between daily mean sodium dietary intake, drinking water sodium, or sodium index (amount of drinking water sodium related to diet sodium intake), and mean blood pressure levels were found . . . Furthermore, levels of sodium in

> drinking water were not related to blood pressure levels or presence of hypertension. (10)

The Iowa Community Drinking Water Survey conducted school and family studies on children and their parents on water sodium and blood pressure levels. Eight Iowa communities with water sodium ranges of 5 to 390 mg/L were surveyed. In the school survey, 2152 children, aged 6 to 13, were compared. Children living in the high sodium towns drank water sodium ranging from 200 to 410 mg/L and those living in the low sodium towns drank water sodium ranging from 4 to 80 mg/L. One third of the homes used home water softeners.

> There was no apparent trend toward increasing blood pressures with increasing sodium levels in the water (with or without home water softeners). (11)

In the family study, 218 children from the school with 204 mothers and 202 fathers were surveyed. This part of the study was designed to compare adult blood pressures in four communities. Two communities had water sodium levels over 200 mg/L, and the other two communities had less than 80 mg/L of water sodium.

> The mean systolic/diastolic blood pressures for families (with or without home water softeners) in the high-sodium communities compared with the low-sodium communities did not vary significantly. (12)

In conclusion, school children and their parents showed little evidence to support an association between blood pressure and sodium in drinking water.

Most of the studies done on water sodium do not support the hypothesis that elevated levels of sodium in our drinking water results in hypertension. Yet, the real question is, "Do higher sodium levels result in higher mortality rates?" That's the bottom line!

People in selected towns in England and Wales were analyzed to see whether sodium rich drinking water correlated with increased mortality. Robertson found:

> . . . no evidence that levels of sodium in drinking water below 250 mg/L lead to any increase in mortality in adults. (13)

Earlier studies done by Schroeder and Sauer support Robertson's findings. In Schroeder's "Relationship Between Mortality from Cardiovascular Disease and Treated Water Supplies," no apparent evidence was observed between sodium levels and death rates. (14)

And in Sauer's work, higher sodium was correlated with lower death rates!

> One certainly needs to be cautious about assuming that a high sodium content of the drinking water is desirable for health in view of the general clinical and laboratory information regarding its deleterious effects in relation to hypertension and congestive heart failure. However, 95 percent of the cities in this study had sodium levels of less than 80 parts per million. It is possible that variations at less than this level do not have an adverse effect upon the risk of dying in the population generally. Another possiblity is that, in the areas with higher sodium content in the drinking water, there may be beneficial factors . . . which offset the effects of sodium. (15)

In Greathouse and Osborne's recent preliminary report on drinking water and cardiovascular disease in the USA, we again find higher sodium levels associated with lower mortality rates.

> . . . sodium means were negatively related to the male and female cardiovascular mortality rates; the associations were statistically significant . . . for both the male and female total cardiovascular-renal and ischemic heart disease mortality rates. (16)

Let's now put the studies and comments on sodium and food and water sodium in perspective and see what it all means. With regards to sodium intake from food, there appears to be substantial evidence to show that by reducing our sodium intake from a range of 8 to 15 grams to a range of 2 to 5 grams daily, we will lower and perhaps, prevent high blood pressure.

Concerning sodium in our drinking water, there is not significant evidence to support the hypothesis that increased levels of water sodium will result in an increased blood pressure. And, more importantly, no study has shown a correlation of elevated water sodium levels to a higher rate of mortality. In fact, some have shown just the opposite. High water sodium levels have been correlated with lower mortality rates!

Yet, the American Heart Association, the EPA and WHO (World Health Organization) believe we should limit, by legal regulation, and/or inform people if the amount of the Na (sodium) in our drinking water exceeds 20 mg/L. (17, 18)

To inform people who may have special health problems of the water sodium levels is one thing and may be useful. However, to regulate the amount of sodium in our drinking water to 20 mg/L may be short sighted. And, perhaps, result in greater health risks!

First, over 40 percent of the drinking water in the U.S. may have sodium levels exceeding 20 mg/L!

> Of 2100 water supplies affecting 50 percent of the U.S. population, which were surveyed from 1963 to 1966 by the U.S. Public Health Service, 42 percent had sodium ion concentration greater than 20 mg/L with approximately 5 percent having levels greater than 250 mg/L. (19)

If we were to follow these organizations' recommendations to limit the amount of sodium to 20 mg/L, we would be in a difficult positon. Because, to accomplish this objective, either municipal water treatment plants would have to remove the excess sodium or it would be up to the individual consumer to do it.

Whether done by the water treatment plant or by the consumer, to remove sodium from drinking water requires using ion-exchange, reverse osmosis or distillation. These methods are effective in removing sodium. But, they would also remove calcium and magnesium (hardness) and in some cases, most of the other minerals as well. As we have seen in the first chapter, hard water is associated with less heart disease. Using distillation and reverse osmosis creates artificial soft water and may also bring on higher death rates.

The central questions to continually ask yourself are, "What constitutes healthy water? What characteristics, elements in the water promote better health and longevity?"

We have seen the value of water hardness. In future chapters, we will see the health value of total dissolved solids and alkaline pH in our drinking water. The problem with trying to remove elevated water sodium is that usually, where you find high levels of sodium, you also have high levels of hardness. The "Preliminary Report on Nationwide Study of Drinking Water and Cardiovascular Diseases" found "that sodium is significantly positively correlated with calcium and hardness."(20) And as earlier mentioned, higher sodium levels had been correlated with lower cardiovascular mortality rates. This anaylsis, plus the Schroeder, Sauer and Robertson studies, do not support the belief that higher levels of water sodium result in higher mortality rates.

This is not to imply that we should ignore the water sodium levels. But, our focus of attention should be on total dissolved solids, hardness and pH. If these are within proper guidelines, then water sodium levels exceeding 20 mg/L should not be cause of alarm for most people. If we want to cut down on our sodium intake, look to our diets. And to water softeners.

Many people have installed water softeners in their homes. This water can be excellent for bathing but can pose serious health concerns.

> A WHO working group proposed that water softened by ion exchange methods should not be used for drinking and for the preparation of food, because the use of home deionized water not only adds sodium but alters the proportion of sodium to potassium, to calcium and to magnesium. Such alterations may be important in the pathogenesis of hypertension. (21)

The usual method of water softening is to add two parts of sodium which pulls out one part of calcium or magnesium. Thus you end up with a water low in hardness and with elevated sodium levels. And that is not a healthy combination! Some methods do not add extra sodium to the water but still result in creating soft water; a water low in minerals. In either case, the water is not optimum for your health.

If we are serious about cutting down on our sodium intake, besides not drinking water processed through water softening devices, our diets offer us lots of opportunities. A recent U.S. survey on salt (sodium chloride) consumption showed an average of 17.2 grams daily. This chart outlines how we consume our daily salt. (22)

(a) 3.2 g (17%) - naturally occuring in food
(b) 5.2 g (33%) - added in cooking or at the table
(c) 8.8 g (50%) - in processed foods

Remember, 90 percent of our sodium intake is from food and 10 percent from drinking water. By eliminating processed food and restricting our habit of salting everything (even before tasting it), we would be able to meet the sodium recommendation mentioned earlier - 5 grams or less. In this case, less is more! Less sodium intake could easily result in fuller health with less hypertension and heart problems.

> We do not know any mechanism for setting threshold levels for chemical carcinogens . . . This has been restated by every single expert committee that's examined the matter up to date. (1)

The Delaney Law states you shall not add any chemical carcinogens to food stuffs. Epstein believes "there is a basis for using this law for water on the grounds that water is not only a food, but also it's a food additive." (2) However, no one has tested this contention in the courts. Perhaps, treating water as a food additive and following the Delaney Law would be a safe way of being confident that the water you drink is not going to lead to cancer.

It is estimated that "60 - 80% of all cancers are environmental in origin." (3)

> There is a growing concensus that the majority of human cancers are caused by chemical carcinogens in the environment and hence, that they are ultimately preventable . . . Several studies have demonstrated the presence of chemical carcinogens in river water, and in treated municipal drinking water, and others have shown that carcinogens are introduced during chlorine treatment. (4)

The amount of chemical compounds discharged in our water, directly or indirectly, is staggering. Not only does surface water contain carcinogenic agents, but so does ground water resources. Even with the EPA drinking water standards, we cannot be assured that the tap water we are drinking is not going to cause us to have caner. Many carcinogenic agents take twenty to thirty years before the effects show up and then, its too late. Regardless of the drinking water standards, each of us are metabolically different and so will react to carcinogenic agents differently. In short, as Epstein warns us, there is no threshold level for chemical carcinogens. We should do everything possible to remove them from our lives.

In later chapters, we will discuss a few of the known carcinogenic agents in detail - fluoridation, asbestos and chlorination. For now, let's look at some fascinating research on water and cancer from a positive, protective perspective. Instead of focusing on the negative harmful properties that may be present in our drinking water, let's look at the beneficial, protective substances in our drinking water.

In 1977, Burton and Cornhill published their paper, "Correlation of Cancer Death Rates with Altitude and with the Quality of Water Supply of the 100 Largest Cities in the United States." Previous research had discovered that with increasing altitude, there was a decrease in cancer rates for countries around the world. In addition to altitude, these authors uncovered a relationship between water quality and cancer. An increase in total dissolved solids, specific conductivity and hardness was found to be correlated with a decrease in cancer mortality. (5)

Specific conductivity is a measure of the ability of water to conduct an electrical charge (valencey as well as concentration is involved). This overall measure is closely related to total dissolved solids. Commonly, about 65% of specific conductivity will give you the amount of total dissolved solids in parts per million. For example, if the specific conductivity is 500 micromhos, the total dissolved solids will be approximately

325 ppm (65% x 500). For our purposes, we will use total dissolved solids.

> The fact that both altitude and the quality of the furnished water supply of these 100 large cities are correlated with the reported deaths must be accepted as proved statistically . . Correlation does not imply causality, since some third factor might conceivably be directly related to both total solids and death rates from cancer . . .
>
> It might be suggested that some carcinogenic agent in trace concentration might be related to total dissolved solids. We think that this is unlikely and that correlating disease in cancer death rates with a high total dissolved solids content is more direct than hidden correlation with a carcinogenic agent . . . (6)

Burton and Cornhill found up to a 20% reduction in cancer deaths in the U.S. cities with the highest total dissolved solids. This is very significant when you consider over 400,000 people in the U.S. die of cancer each year! Drinking water high in TDS could result in saving 80,000 lives per year from cancer deaths.

In 1980, Burton and Cornhill published an extension of their first study entitled, "Protection from Cancer by 'Silica' in the Water Supply of U.S. Cities." Instead of looking for environmental factors that are carcinogenic the authors looked for specific agents in water that could reduce cancer rates.

In their earlier study, Burton and Cornhill found:

> . . . remarkable significant reductions of cancer (all sites) with altitude, with concentration

> of total dissolved solids, with electrical-conductivity (which is closely related to total solids) and with pH of the finished water supply . . . Correlation was significant for all of the major ions, e.g. Na^+, Ca^{2+}, Mg^{2+}, HCO_3^-, SO_4^{2-} and Cl^-. All had negative correlation coefficients, i.e. the greater the concentration of these ions, the less the death rate from cancer. The only positive correlation was with H^+ ions (negative with pH). (7)

The authors' comments on pH are confusing. In their first study, "the correlation with pH (was) not statistically significant . . . but the correlations for hardness, dissolved solids and specific conductance (were) astonishingly significant." (8)

However, in their second study, in which they did not specifically analyze pH, but reviewed their earlier findings, they state that "the only positive correlation was with H^+ ions (negative with pH)." (9) That is, the more acid or negative the water is, the higher the cancer rates. They later go on to recommend drinking water should have a pH of 8 or greater.

Very few research studies have looked at the positive or negative effects of the pH of drinking water and its effects on the body. When we repeatedly see studies showing the beneficial significance of total dissolved solids and hardness, we are on more solid ground to make intelligent, practical recommendations.

However, Burton's remarks on pH are very similar to Schroeder's findings. Schroeder found that an alkaline pH resulted in less cardiovascular diseases than water with an acid pH.

> The pH of the finished water varied widely from city to city, from an extreme low of

3.9 . . . to an extreme high of 10.4 . . . The correlation was again negative; i.e., alkaline waters appeared to 'protect' at about the 1% level of significance. (10)

Limited studies have been done on the effect of drinking water's pH and digestion. There are some indications that people with low stomach acidity seem to be helped with drinking water that is more alkaline. And, those with hyperacidity seem to be helped with drinking water that is more acid. Why? Drinking a more alkaline water at mealtime seems to hold the food in the stomach longer, giving the gastric juices a longer time to assist in digestion. Whereas, an acid water causes the stomach to empty its food quicker, thus cutting down on the amount of acid naturally produced by the stomach. Carbonated water is usually acidic and results in more rapid emptying of the stomach and less hyperacidity. (11, 12)

In these studies, the effects of carbonation have been attributed to the presence of carbon dioxide. However, there is good reason to believe that a similar effect on stomach emptying and acidity can be achieved by using water with the proper pH.

The pH also affects the corrosive action of water. Corrosive water has a tendency to leach from pipes harmful substances such as lead and cadmium which end up in our drinking water. In the past, it was thought that soft water was corrosive and hard water was not. However, careful research from the Netherlands shows that the key factor to determine whether or not a water is corrosive is the pH and not whether the water is soft or hard. (3)

Your water pipes may contain lead or cadmium solders. An acid or low pH water (below 7.0) could result in leaching these harmful agents into your drinking water. If the pH is neutral (7.0) or alkaline (higher than 7.0), your drinking water should be okay. But, as a safety measure, flush your faucets for thirty seconds each morning before drinking. This will help eliminate the build-up of lead and cadmium that may be present.

The research on water pH is limited to just a few studies. Yet, it appears adequate to recommend that under normal circumstances, drinking water should have an alkaline pH.

In their latest study, Burton and Cornhill analyzed forty constituents appearing in water and correlated this with cancer rates. Their findings were:

> No significant evidence of carcinogenic agents (other than H+) results, but an unexpected protection by 'Si02' was found. (14)

> . . . The most probable mechanism is by stimulation of immunodefences, initiated by effects on phagocytic cells. The taking of hydrogen ions out of dissociation by silicia acids and silicates may also contribute. (15)

For a fuller explanation of the possible protective mechanism of silica and silicon compounds, the reader is invited to read the complete report. The authors end their report with this challenge:

> If the control of cancer were the dominant consideration, and the (water) concentration were made of total dissolved solids to 300 ppm or more (or electrical conductivity to 500 micromhos), along with alkaline pH, e.g. greater than 8, and minimum concentration of 15 ppm of silica, there would be a substantial reduction of death rates from cancer in these

> cities, which have a third of the population of the U.S.A. . . There is a high probability that it would exceed 10% resulting from increasing Si02 alone. Our estimate would be as high as 25% . . . (16)

The authors emphasize that this type of water is now being drunk in several cities and is not experimental or hazardous.

> The 'clinical investigation' has already been made, in the several cities that already have the suggested high levels of conductivity, pH and Si02, so the proposed guidelines do not present any unknown hazard to the population of the cities. (17)

By way of support, Sauer also found correlations with total dissolved solids, hardness and silica. (See Tables 1, 2 and 3 in Chapter One). A beneficial effect in lessening heart disease and cancer was found when these substances were present in drinking water.

The correlation of silica and cancer was especially evident in Sauer's study. "The highest negative correlations are with cancer, and silica rather consistently shows a higher correlation with cancer than do any of the other substances considered." (18) In short, the more silica in the drinking water, the less cancer.

Whether or not Si02 (silica and silicon compounds), will be found to be significant in other studies is impossible to predict. Based on the research on heart disease, it seems different minerals are found to be significant in different studies, in different geographical locations. In some studies, calcium is highly correlated, in others magnesium, in others cadmium, and so on. The more we try to localize the element, the more inconsistent the results. The physiological requirements and effects of a specific mineral in the drinking water may be different from one location to another. With this perspective, we can see the true value of all minerals even though the importance of one specific mineral may vary with the location.

This point is highlighted in the study in Seneca County, New York on cancer mortality. Seneca County has one of the lowest mortality rates for all cancer sites in New York.

> The combined mortality rates for all high-rate neoplasms and for all malignant neoplasms combined are considerably lower in Seneca County, New York, than in its surrounding counties.(19)

The drinking water of Seneca County is unique compared to the surrounding counties and New York itself. For example, the selenium concentration in its water is about 1.6 times the average concentration in New York. In addition to the selenium, the drinking water is high in potassium and salinity (salts).

> Prevalence and availability of selenium, high salinity of the water and especially high concentrations of potassium cations may belong to the anticarcinogenic factors contributing to the low rates. (20)

In Seneca, silica was not specifically studied, although due to the high salinity, it could be present. This report shows that in different regions, different trace minerals may be very important in cancer protection and prevention.

Thus far, we have seen the importance and benefits of total dissolved solids and specific trace elements in relation to cancer. Before ending this chapter, let's also consider water hardness.

Both Sauer and Burton found a negative correlation for hardness and cancer. Although these two researchers focused in on total dissolved solids, they also found that hard water was associated with lower cancer rates.

The same observation was found in Quebec. In "Cancer and the Physicochemical Quality of Drinking Water in Quebec," the authors state:

> Generally speaking, the softer the water, the higher the incidence of neoplasms of the rectum, other organs of the digestive system and prostate. (21)

The overall pattern that emerges from the studies on heart disease and cancer and their relation to drinking water is this: Drink water that is high in hardness and high in total dissolved solids and with an alkaline pH. If you do, you will be protected by the various water factors which act individually and synergistically. These factors either protect you from toxic properties and/or give you nutrients in a form that your body can accept and are significant in maintaining good health.

Now, let's turn our attention away from the protective elements in water and look at some of the more prevalent harmful carcinogenic agents in our drinking water.

More people have died in the last thirty years from cancer connected with fluoridation than all the military deaths in the entire history of the United States.(1)

Dr. Dean Burk has spent more than fifty years in cancer research. Many of those years were with the National Cancer Institute. He began studying the association between artificially fluoridated drinking water and cancer in 1975 and, along with Dr. John Yiamouyiannis, has published several studies. One of their studies found the cancer death rates of large U.S. cities, having fluoridated water since the 1950's, exceeded those in cities where fluoridation had not been started or had started much later.(2)

Fluoridated Cities Higher Cancer Rates	Non-Fluoridated Cities Lower Cancer Rates
Chicago	Los Angeles
Philadelphia	Boston
Baltimore	New Orleans
Cleveland	Seattle
Washington, D.C.	Cincinnati
St. Louis	Atlanta
San Francisco	Kansas City
Milwaukee	Columbus
Pittsburgh	Newark
Buffalo	Portland, OR

The Burk-Yiamouyiannis study has been the subject of much debate. However, court cases on fluoridation in Pennsylvania and Illinois, after hearing expert testimony, have up-held the validity of this study and it's findings. In the 1982 Illinois fluoridation case, after a forty-day trail of testimony, Judge Ronald A. Niemann ruled, "a conclusion that fluoride is a safe and effective means of promoting dental health cannot be supported by this record." (3)

If you are drinking artificially fluoridated water, your chances of dying of cancer are increased five to fifteen percent. (4)

Several studies in fluoride toxicity have been linked to genetic damage, birth defects and cancer.

Fluoride injures the genetic material of the cells in both plants and animals. Fluoride inhibits DNA repair enzymes. This slow down in DNA repair allows damaged cells to replicate themselves which leads to increased chromosomal damage. "There is a distinct possibility," according to Dr. Yiamouyiannis, "that fluoride interferes directly with the production of deoxynucleotide, the building block of DNA. If this interference occurs, chromosomal damage and cancer can surely result." (5)

This chromosomal interference is apparent in monogolism or Down's syndrome. After reviewing the literature linking fluoride to Down's syndrome, Dr. Waldbott writes:

> In summary, with exception of the National Intelligence Surveillance Survey, which was based on admittedly incomplete ascertainment in only five major cities, all large-scale U.S. studies to date have shown higher incidence or prevalence rates of Down's syndrome births in communities with elevated levels of fluoride in the drinking water. (6)

Chronic fluoride toxicity has a multitude of symptoms. In Dr. Waldbott's book, Fluoridation: The Great Dilemma, he recommends a test to detect fluoride intoxication. The following procedures are to be rigorously followed:

> Avoid all fluoridated water (substitute distilled or other nonfluoridated, low-fluoride water), fluoridated beverages, fluoride-rich foods (tea, ocean fish, gelatin, skin of chicken, etc.), fluoridated toothpastes, and any other source of environmental fluoride, including cigarette smoke and industrial pollution. If symptoms are in fact caused by fluoride, they should diminish markedly within a week and largely disappear within several weeks. If symptoms persist, consult a physician for possible alternative problems. True fluoride toxicosis can be reproduced by re-exposure to fluorides from whatever source.(7)

According to Dr. Waldbott, persons suffering from fluoride poisoning may exhibit the following symptoms:

- Chronic fatigue not relieved by extra sleep or rest
- Headaches
- Dryness of the throat and excessive water consumption
- Frequent need to urinate
- Urinary tract irritation
- Aches and stiffness in muscles/bones (arthritic-like pain)
    - In lower back
    - In neck area
    - In jaws
    - In arms, shoulders, legs
- Muscular weakness
- Muscle spasms (involuntary twitching)

Tingling sensations in fingers (especially) and feet

Gastrointestinal disturbances

Abdominal pains	Blood in stools
Diarrhea	Bloated feeling (gas)
Constipation	Tenderness in stomach area

Feeling of nausea (flu-like symptoms)

Pinkish-red or bluish-red spots (like bruises, but round or oval) on the skin that fade and clear up in 7 to 10 days

Skin rash or itching, especially after showers or bathing

Mouth sores (also from fluoridated toothpaste)

Loss of mental acuity and ability to concentrate

Depression

Excessive nervousness

Dizziness

Tendency to lose balance

Visual disturbances

Temporary blind spots in field of vision

Diminished ability to focus (possible retinal damage) (8)

If all of these problems can result from too much fluoride, why do we have fluoridation? A good question and a difficult one to truthfully answer. The main reason given by pro-fluoridationists is that up to 1 ppm of fluoride will reduce cavities. No geniune scientific research has supported this contention.

Although attempts have been made, the United States Center for Disease Control and the British Ministry of Health admit that no laboratory experiment has ever shown that 1 ppm fluoride in the drinking water is effective in reducing tooth decay. Furthermore, they admit that

> there are no epidemiological studies on humans showing that fluoridation reduces tooth decay that would meet the minimum requirements of scientific objectivity. (9)

Dr. Burk states "that the most that can be given to the fluoridation argument is that up to the age of puberty there may be a temporary delay of about one tooth decay." (10)

Dental health is directly related to our diet and nutritional intake. The clearest evidence of the relationship of diet to tooth condition comes from Dr. Weston Price's book, Nutrition and Physical Degeneration. Dr. Price, a dentist, traveled around the world and studied fourteen primitive races in various stages of modernization. He found that as the primitive people left their native diets and adopted the diets of the white, civilized races, their dental health showed significant decline.

> Whereas the average incidence of dental caries in the isolated primitive groups regardless of location or tribal characteristics involved, on the average, is approximately one tooth per hundred teeth examined, within a relatively short time after contact was made with the white race, this factor increased to an average of thirty teeth of each hundred teeth. (11)

Basically, the white races' diet consisted of refined sugar and white flour processed products from food grown with heavy reliance on pesticides. Whereas the native diets consisted of whole food grown or raised by natural means.

Let's face it, dental cavities are not a fluoride deficiency disease!

However, when we look at the following examples we can see how the quality of drinking water can give us a clue to understanding how fluoride was mistakenly associated with dental health.

In 1942, Hereford, Texas was heralded as a "town without a toothache". The drinking water had 2.3 - 3.2 ppm of natural fluoride but it also have generous amounts of calcium, magnesium and other minerals. Pro-fluoridationists pointed to the fluoride levels as the reason for low levels of dental problems. But, Dr. G.W. Heard, Hereford dentist for thirty-five years, called attention to the hard drinking water as a possible reason for the benefits. "The damage fluoride does is far greater than the good it may appear to accomplish," stated Dr. Heard. He goes on to report, fluoride "makes the teeth so brittle and crumbly that they can be treated only with difficulty, if at all."(12) It is a well known and accepted fact that even low fluoride levels can lead to mottled teeth.

Dr. Heard's observation, noting the importance of water hardness and the negative effect of fluoride, is significant.

Colorado Springs has a high level of fluoride in its drinking water (2.5 ppm) and yet has a high level of cavities. Colorado Springs drinking water is exceptionally soft (low in calcium and magnesium) whereas the water in Hereford is high in these elements.

The connection of the effect of hard and soft drinking water on teeth was also made by Dr. Yiamouyiannis in the study comparing the cities of Rockford, Illinois (non-fluoridated) with Aurora, Illinois (naturally fluoridated). "They found that people who lived in hard water areas had better teeth than people in soft water areas." (13) Whether the water was fluoridated or not fluoridated was not important to dental health.

Is it possible that the hardness of drinking water may provide protection against elements in the water that could be injurious to health? Let's look at an experiment done by Dr. Yiamouyiannis on fluoride toxicity in trout. This quote is from his testimony in the Third Judicial Circuit Court of Illinois.

> I tested a theory to see if water hardness could counter-act low levels of fluoride in trout and we confirmed that water with higher calcium and magnesium levels were able to counter-act the toxic effects of fluoride. The significance of this (is that) in a soft water system, fluoride toxicity on biological systems would be greater than it would be in hard waters and most of your naturally fluoridated areas are hard water areas . . when you fluoridate soft water systems, you would, on top of the normal toxic effects of natural fluoridation, get additional effects due to the fact that you had no calcium and magnesium ions to offset the toxicity of fluoride. (14)

This work was substantiated by Newhold at Utah State University.(15) "The softer the water, the more fluoride passes through the intestinal wall."(16)

These are very significant observations because of two reasons: First, it shows that high fluoride levels - artificial or natural - can be harmful. Second, and more importantly, it shows that the key to how fluoride will react in the body is based on the hardness of the water. If the water is hard, the harmful effects of high fluoride levels are minimized. If the water is soft, the harmful effects are amplified. The towns of Hereford and Colorado Springs support this interpretation.

An interesting study would be to re-analyze the cities listed earlier with the high cancer rates and fluoridation and analyze the amount of total dissolved solids, hardness, and pH.

Unfortunately, the fluoridation issue is usually discussed in highly emotional and irrational presentations - written and verbal. And can lead to misinformed decisions with tragic consequences.

The clinical and experimental research is best summarized by the ruling of Judge Niemann quoted earlier and here paraphrased - fluoride is not a safe, nor an effective dental health measure.

Our investigation of the fluoridation record world-wide shows a revealing pattern. Most countries have not adopted fluoridation. Those that did have changed their policy and discontinued its use due to its ineffectiveness and/or dangers.

TABLE 5

WORLD-WIDE FLUORIDATION SCORECARD (17)

Austria 7,500,000	No Fluoridation	"Will not be carried out."
Belgium 9,750,000	No Fluoridation	Only one small experimental plant, now discontinued.
Denmark 5,000,000	No Fluoridation	Forbidden, by law in food and water.
Egypt 37,230,000	No Fluoridation	U.S. pressure to fluoridate rejected.
France 52,000,000	No Fluoridation	Never considered essential to good health.
Germany 61,000,000	No Fluoridation	Discontinued in 1971 after 18 years of experiments - For health & legal reasons.

Greece 9,000,000	No Fluoridation	No experimental programs have every been introduced.
Holland 13,500,000	No Fluoridation	Discontinued in 1976 after 23 years of experiments involving 9,800,000. On August 31, 1976, by Royal Decree, all permissions to fluoridate were cancelled.
India 598,170,000	No Fluoridation	Endemic fluorosis occurs with varying intensity in many parts of India. The removal of fluoride from water is a major public health problem. Defluoridation units are functioning in parts of India.
Italy 55,000,000	No Fluoridation	In some areas, public drinking water supplies are defluoridated.
Luxembourg 360,000	No Fluoridation	
Norway 4,000,000	No Fluoridation	Legislation designed to make fluoridation compulsory was rejected by the Norwegian Parliament in 1975.
Spain 35,000,000	No Fluoridation	Forbidden by law.

Sweden 8,200,000	No Fluoridation	Forbidden by law. Discontinued in 1969 after a 10 year experimental program. The WHO was asked to produce evidence to support its earlier claim that fluoridation was 'safe'. No evidence was produced. Parliment declared fluoridation illegal on Nov. 18, 1971.
Switzerland 6,500,000	Limited Fluoridation	One experimental program only since 1959, and it serves just 4% of the total population. In December, 1975, the Swiss Health Dept advised the cessation of fluoridation in Switzerland: "Due to its ineffectiveness".
Portugal 8,500,000	Limited Fluoridation	One small experimental plant still in operation.
Finland 4,700,000	Limited Fluoridation	One small experimental plant operating since 1959, and serves only 1½% of the total population.
Chile	No Fluoridation	In 1977, stopped fluoridating.
Quebec	No Fluoridation	In 1979, the government of the Province of Quebec dropped the bill for mandatory fluoridation, after a 5 year study by scientists from ten different fields -

		government will no longer fund floridation.
Wales	No Fluoridation	In 1982, according to news report, the Wales government is not extending fluoridation due to the Water Works Association's insistence on legal and financial indemnity against suits that may be filed in the future by people harmed by fluoridation.

Based on the research presented on heart disease and cancer with water quality, namely total dissolved solids, hardness and alkaline pH, we would expect to see that fluoride in and of itself may not be the only culprit. The kind of water that it is found in greatly amplifies the problem. This statement is not intended to minimize the dangers of fluoridation, but to put it in a context that gives us a more useful understanding of the beneficial elements in drinking water.

Fluoridated drinking water should be avoided. The sodium fluoride used is a poison and we don't need additional poisons in our life. If you want healthy teeth, correct your diet and don't add poisons to your water. A thorough review of the history, research and politics of fluoride can be read in Fluoride: The Aging Factor by Dr. Yiamouyiannis.

In Dr. Burk's opinion, "one tenth of all cancer deaths in this country can be shown to be linked to fluoridation of public drinking water. That comes to about 40,000 extra cancer deaths a year. It's ten times the number who die from asbestos induced cancers and exceeds deaths from breast cancer." (18)

Yet, 40% of the people in this country continue to drink fluoridated water. (19)

5 • ASBESTOS AND CANCER

Based on the extensive research by Dr. Irving J. Selikoff and others, we now have well-documented relationships between occupational asbestos exposures and increased respiratory and digestive cancers in exposed workers.(1) The first hint of the possible harm of asbestos inhalation was noticed by Dr. Selikoff in 1924 and his research to date has presented convincing evidence on the relationships between asbestos exposure and high rates of cancer. Although this work deals with the inhalation of asbestos fibers and not fibers in drinking water, the research presents lessons and principles that are applicable to harmful and deadly substances in our drinking water.

In 1954, cancers of the lung were being described among asbestos workers. This observation started a review of health records to see if there was any relationship between cancer and asbestos.

> For 30 years, laboratories tried to produce cancer in animals with asbestos and weren't able to. We learned in 1963-4, and now every pathologist can produce them with ease, but for 30 years we couldn't.(2)

This lack of ability to detect the cancerous effect of asbestos on animals for 30 years is amazing and frightening. Most of our tests and subsequent standards as to whether a substance is harmful to humans and in what dose, is first based on animal studies. Clearly, animal studies may not be as reliable as we would like. This example highlights how cautious we need to be in exposing ourselves to new substances.

In addition, Selikoff points out the effects of asbestos takes a long time to manifest. He calls it the twenty to thirty year rule for environmental disease. This time factor was demonstrated in his study of several hundred, and later several thousands of asbestos workers. He found workers with less than twenty years exposure had mostly normal X-rays, despite the fact that they worked with asbestos fibers almost every day. However, after twenty years, the X-rays tended to become abnormal. Not infrequently, they were extensively abnormal.(3)

Selikoff's work also revealed the simultaneous interaction of two harmful agents - smoking and asbestos. The interaction of these two carcinogenic influences produce what Selikoff called the "multiplying effect". He found that the high rate of lung cancer among asbestos workers (one out of every five) was not due to just smoking or just asbestos exposure but to the interaction of both substances on the workers' lungs.

From this research, we see three important principles or lessons:

1. Animal studies may not reveal carcinogenic agents.
2. It takes twenty to thirty years of exposure to produce cancer and when it is detected it can be very advanced.
3. The interaction of two or more carcinogenic or weakening agents leads to greater problems than one carcinogenic agent alone.

We should keep these findings in mind as we are constantly being told of the minimum risk a substance has or is going to have on our health. The interaction of a variety of carcinogenic agents in very low doses creates unknown synergistic effects which may prove deadly. Generally, when we test animals to set water standards for a variety of chemicals, we test them for one substance at a time and do not study, nor know the synergistic effect of other substances present in the water. Keep in mind Epstein's warning, "We do not know of any mechanism for setting threshold levels for chemical carcinogens."(4) Synergism greatly complicates this task!

With this background, let's now look at the research on asbestos and drinking water. Perhaps, the best known incident of asbestos and drinking water is the case of Duluth, Minnesota. Since 1955, Reserve Mining Company has been dumping taconite wastes (an asbestos fiber) into Lake Superior. In Sigurdson's study, "Cancer Morbidity Investigations: Lessons from the Duluth Study of Possible Effects of Asbestos in Drinking Water", cancer death rates and sites of Duluth residents were compared to residents of Minneapolis - St. Paul. The authors point out that they believe sufficient time has not elapsed to determine the full associations of asbestos and cancer. Residents have been exposed to high levels of asbestos in the water for ten to fifteen years and it may take twenty to thirty years, as in asbestos inhalation to clearly and unequivocally show the impact of asbestos.

Even with this relatively short time exposure, the authors found:

> Mortality rates in Duluth are substantially greater than mortality rates in Minneapolis for the following sites: stomach (male and female), small intestine (male and female), rectum/rectosigmoid (male and female), pancreas (female), total GI sites (female) and lung (male). However, the corresponding incidence rates for both cities are quite similar.(5)

From this we can see that even though the exposure to asbestos has been for only ten to fifteen years, we are starting to see that where the people contact cancer is related to the findings of asbestos inhalation. Although the overall current mortality is similar between the two regions, the location of cancer indicates that asbestos is showing its effect.

In Iowa City, Iowa, asbestos fibers in drinking water have been linked with stomach cancer at a rate three times higher than people not exposed.(6) In a six year study in the San Francisco Bay Area, the relationship between the presence of asbestos fibers in drinking water and the incidence of selected cancers in the population served by the water supply showed the following:

> Significant relationship between chrystile asbestos content of drinking water and white male digestive tract, esophageal, stomach and pancreatic cancers. For white females, significant relationships were found for esophageal, stomach, digestive related organs and pancreatic cancers. These associations appear to be independent of income, education, asbestos occupation, martial status and mobility. (7)

The discovery of a link between cancer and asbestos inhalation is beginning to be seen in the relationship between asbestos and drinking water. Let's hope we don't wait thirty years to realize that asbestos fibers don't belong in our drinking water!

6 • CHLORINATION: A LINK BETWEEN HEART DISEASE AND CANCER

> The origin of heart disease is akin to the origin of cancer. (1)

The use of chlorine in the United States to 'purify' water supplies began in the late 1890's and gained relatively wide acceptance by 1920. Dr. Joseph Price believes there is a definite correlation between the introduction and widespread application of chlorination of water supplies and increasing incidence of heart attacks. He writes:

> It is, in my opinion, certainly one of the greatest paradoxes of recorded history that one of the very same public health measures which have been primarily responsible for the great increase in statistical life expectancy in the Western world should also unsuspectedly be responsible for many of the chronic disorders of later life. (2)

Dr. Price maintains in atherogenesis the primary agent is chlorine. It is only one cause, but it is the essential cause.

> Nothing can negate the incontrovertible fact that the basic cause of atherosclerosis and resulting entities such as heart attacks and the most common forms of stroke is CHLORINE - the chlorine contained in processed drinking water! (3)

For a poignant analysis and demonstration of the connection between heart disease and chlorine, a lengthy summary of Dr. Price's description of how the heart functions and the impact chlorine has is presented.

> The heart, like any other muscle in the body, must constantly receive a flow of blood to it carrying food and oxygen in order to live . . . The heart muscle itself cannot use any of the blood contained within its pumping chambers. Instead the supply of blood to the heart muscle comes entirely from two little vessels, called coronary (heart) arteries, which arise from the main artery of the body, the aorta, run along the outside surface of the heart and then enter into the heart muscle to feed it.
>
> Since these two little arteries are the only source of blood to the heart muscle, when one of them (or one of their branches) suddenly becomes blocked off, the portion of the heart muscle supplied by the involved artery dies.(4)

When this happens, we have a 'heart attack'. Now, what causes this blocking off or occlusion?

> While the actual total occlusion of a coronary artery which results in an acute heart attack takes place suddenly, it only occurs in previously diseased arteries affected by a pathological process known as atherosclerosis. (5)

Atherosclerosis or hardening of the arteries is characterized by the gradual narrowing of the arterial wall. The opening becomes smaller and smaller. This narrowing is due to plaque formation.

> A heart attack occurs when spontaneous bleeding in or around an atheromatous plaque causes a clot to form through a coronary artery . . . There could not be a heart attack if the coronary arteries were not partially closed by artheromatous plaque. (6)

From this, we can see that a heart attack is the end result of a more general disease process in which deposits are built up on the insides of the arteries. "The infiltration of the artery walls begins in most cases at least ten to twenty years before any overt symptoms are evident." (7)

> When the arteries feeding the brain are affected by atherosclerosis the stage is set for the occurance of the other great killer . . . the stroke.(8)

The list goes on to include senility of old age and male sexual impotency both due to vascular insufficiency.

To experimentally test his thesis on the relationship of chlorine and atherosclerosis, Dr. Price conducted animal experiments. We will discuss these experiments at length in the next chapter, <u>The Water Story and Animal Experiments</u>. Chlorine caused atherosclerosis in 95% of the animals!

Price makes a very insightful comment on the possible interaction of chlorinated water and hard water.

> The documented lower incidence of coronary heart disease in areas with hard water could possibly be explained by postulating chemical reactions between free chlorine, an extremely active chemical, and the ions which cause hardness of water resulting in biologicaly innocuous chlorides. (This is not to exclude the possibility

of some other, more complex biological mode of action of the hard water ions. (9)

This possible protective aspect of hard water reminds us of the several studies cited on hard water and heart disease, and total dissolved solids and cancer.

While Dr. Price does a thorough job of demonstrating the relationship of chlorine to heart disease, he does not consider chlorine and its relationship to cancer.

To understand how chlorine is related to cancer, we need to understand what chlorine is and how it interacts with other chemicals and constituents that naturally occur in water.

In the chlorination process itself, chlorine combines with natural organic matter, such as decaying vegetation, humus, to form potent cancer-causing trihalomethanes or haloforms. Trihalomethanes collectively include such carcinogens as chloroform, bromoform, carbon tetrachloride and bischloroethane and others.

> . . . Certain substances in chlorinated water . . . have proved carcinogenic in laboratory animals. Prominent among these substances are the trihalogenated methanes, including chloroform and bromoform, which are sometimes present in considerable amounts, both in natural waters and in chlorinated water. In naturally collected water, the carcinogens may derive from plants such as bracken. Organic materials may be leached from the soil, possibly humic acids, which will react with chlorine to produce the suspected carcinogens. (10)

The amount of trihalomethanes allowed in our drinking water is regulated by the EPA. Although the maximum amount allowed by law is 100 ppb (parts per billion), "a 1976 study showed that 31 of 112 municipal water systems surveyed exceeded this limit." (11)

Let's look at chloroform, a known poison, more closely.

> Chloroform . . . is a known animal carcinogen; it is present in measurable quantities in nearly all chlorinated municipal drinking water systems; and it is a known by-product of water chlorination. (12)

To answer the question of how widespread are trihalomethanes (THMs) or haloforms in our drinking water, the National Organics Reconnaissance Survey (NORS) measured the concentrations of six halogenated compounds in the finished water supplies of eight U.S. cities.

> Water samples were collected from all 10 EPA regions and analyzed for chloroform (CHCL3), dichlorobromomethane (CHBrCl2), dibromochloromethane (CHClBr2), bromoform (CHBr3), carbon tetrachloride (CCl4), and 1,2-dichloroethane (C2H4Cl2). Quantitative results demonstrated that trihalomethanes, particularly chloroform, occur ubiquitously in finished drinking water and that THMs are practically nonexistent in raw water sources. (13)

Not only is chloroform everywhere in our drinking water but the levels are very high. And, the cause is not primarily industrial pollution!

> Quantitative results of the EPA surveys also indicate that chloroform occurs at concentrations far exceeding levels of other THM's and non-THM organic contaminants, including those

of industrial origin. In other words, waterborne chloroform is primarily a result of chlorination, not industrial discharge. (14)

By 1975, the number of chemical contaminants found in finished drinking water exceeded 300; roughly half are halogenated. (15)

Wilkin's study lists the following chemical carcinogens found in drinking water.

TABLE 6

Chemical Carcinogens Found in Drinking Water (16)

Compound	Highest Concentration in Finished Water (1977) *
Human carcinogen	
Vinyl chloride	10
Suspected human carcinogens	
Benzene	10
Benzo (a) pyrene	D+
Animal carcinogens	
Dieldrin	8
Heptachlor	D+
Chlordane	0.01
DDT	D+
Lindane	0.01
β-BHC	D+
PCB (Aroclor 1260)	3
Chloroform	366
α-BHC	D+
Carbon tetrachloride	5

Compound	Highest Concentration in Finished Water (1977) *
Animals carcinogens (Cont.)	
Trichloroethylene	0.05
Diphenylhydrazine	1
Aldrin	D+
Suspected animal carcinogens	
Bis (2-chloroethyl) ether	0.42
Endrin	0.08
Heptachlor epoxide	D+

* Concentrations in μg/liter
D+ = Detected but not quantified

It is impossible to tell how many additional carcinogenic agents are created from the interaction of chlorine in water with these chemicals.

Recently, ABC News aired a revealing and frightening show on drinking water - "Water - A Clear and Present Danger." They point out that:

> Over 700 chemicals have already been found in our drinking water. The EPA has targeted 129 as posing the greatest threat in water, but to date the EPA requires that our drinking water be tested for only 14 of those 129. (17)

For example, in New Orleans, Louisiana which takes its tap water from the Mississippi River, there has been found "sixty-six new carcinogenic compounds created in Mississippi drinking water when chlorine combines with methanol, carbon disulfide and others."(18) It's no wonder that an analysis of cancer mortality in Louisiana, from people drinking water from the Mississippi River, shows higher rates of total cancer, and specifically sites cancer of the urinary organs and gastrointestinal tract. (19)

In 1981, additional research was published on the incidences of colon and rectum cancer in relation to the drinking water in Louisiana. Colon cancer, which has usually been associated with cultural and dietary factors, did not show any significant relationship to the water quality variables. However, rectal cancer was strongly associated with Mississippi River surface water and chlorination. In fact, the closer one lives to the mouth of the Mississippi River, the higher the rate of rectal cancer due to the higher levels of chlorinated surface water. (20)

In Eric County, New York,

> . . . research has revealed positive associations between surface water as a water supply source and the incidence of esophageal and pancreatic cancer, and between THM concentrations and the incidence of pancreatic cancer among white males. (21)

In Washington County, Maryland, the vital records of 31,000 subjects were analyzed to study the effect of their source of drinking water to cancer rates. Incidence rates of cancer of the bladder among men and for cancer of the liver among women were nearly two fold higher in those whose water source was chlorinated surface drinking water, as opposed to those who drank unchlorinated ground water. In addition, a complementary mortality study also suggested an association of chlorinated water with cancer of the liver and urinary tract.(22) Because of the relatively small number of subjects, the results were not statistically significant. However, the scientific research linking chlorination with increase cancer incidences continues to grow.

The bladder cancer results reported here are therefore consistent with recent studies that used different drinking water exposure data and were performed in other geographic regions of the country . . . The observed association between cancer of the bladder and drinking water is not unique to Washington County. Kuzma et al. reported elevated white male bladder cncer mortality rates in Ohio counties relying on surface water supplies. Hogan et al. observed a 'weakly suggestive' association between bladder cancer mortality in white females and CHCL3 concentration data derived from the US EPA Region V Survey. Cantor et al. reported an association between bladder cancer and a trihalomethane exposure index.(23)

"Chlorine is so dangerous," according to biological chemist, Dr. Herbert Schwartz, "that it should be banned."

Putting chlorine in the water supply is like starting a time bomb. Cancer, heart trouble, premature senility - both mental and physical - are conditions attributable to chlorine - treated water supplies. It is making us grow old before our time by producing symptoms of aging such as hardening of the arteries. Tests have shown that chlorine does have devastating effects on such living organism as plant seeds. When it is present in the human body, you may expect a premature end to cell life and death. It has been shown that where people drink mountain water, pure and free of the chlorine found in big-city water, they tend to live longer. I believe if chlorine were now proposed for the first time to be used in drinking water, it would be banned by the Food and Drug Administration. (24)

Drinking tap water that is chlorinated is hazardous if not deadly to your health. It's as simple as that!

More and more research is being published on the harmful effects of chlorine. A recent article in the American Journal of Public Health highlights this point. "Chlorine is currently being re-evaluated as the standard for disinfection of drinking water and waste water."(25)

Alternatives to chlorine are: chlorine dioxide, ozone, bromine chloride, ultraviolet light and ultrasonics. A useful and informative chart on the advantages and disadvantages of these methods follows:

TABLE 7

Alternative Disinfectants - Advantages and Disadvantages (26)

Chlorine Dioxide

Advantages

Effective against many microbes.
More effective than chlorine over short contact.
Strong oxidant, long residual.
Good taste, odor, color control.
Iron, manganese removal.
Not reactive with ammonia or aromatic organics to yield trihalomethanes.
Forms chlorinated organics less readily than chlorine.

Disadvantages

Cost.
Chlorine, chlorite, chlorate are formed in production.

Chlorine Dioxide

Disadvantages (Continued)

With excess chlorine, trihalomethanes are formed.
Chlorite and chlorate oxidize hemoglobin. Chlorite is a hemolytic agent.
More data on acute and chronic effects of the production by-products needed.

Ozone

Advantages

Strong oxidizing agent.
Good color, taste, odor control.
No trihalomethanes formed. Can oxidize trihalomethane precursors.
With U.V. can remove pesticides, PCB's (high concentrations and contact time needed).
Effective against a variety of microbes.
Improves flocculation and settling.

Disadvantages

No residual effect.
Organic reaction products largely unknown.
Epidemiology of ozone effects in potable water not available.

Bromine Chloride

Advantages

All advantages of chlorine.
More reactive than chlorine on microbes.
Bromamines formed are more effective than chloramines for microbe removal.

Bromine Chloride

Disadvantages

All disadvantages of chlorine.
Brominated organics formed generally more toxic than chlorinated organics - but are more unstable.
More data needed on environmental effects.

Ultraviolet Light

Advantages

Effective against many microbe types.
No chemical by-products or toxics.

Disadvantages

Penetration capacity through water limited.
Color, turbidity, organics can reduce potential.
UV harmful to eyes, skin.
No residual effect.

Ultrasonics

Advantages

Effective against many microbe types.
Increases settling rate of activated sludge and mixed liquor.
Aids in hardness removal.

Disadvantages

Thick films of water attenuate sound and reduce effectiveness.
Cost.

As you can see from a close reading of this table the alternatives to chlorine are not without their health dangers. Potentially ozone and ultraviolet light may offer the best solution to the water disinfection. Most treatment plants that are trying to cut down on the use of chlorine are using chlorine dioxide and/or bromine chlorine which have most of the dangers of chlorine! Let the consumer beware of these chemical disinfecting agents. If they are in your water, you should have them removed with the proper home filtration units.

Before ending this chapter on chlorination, let's discuss a reasonable explanation and link between these two degenerative diseases: heart disease and cancer. In Dr. Price's work, Coronaries/Cholesterol/Chlorine he states chlorine is the essential cause of atherosclerosis but is unable to explain the mechanism of how chlorine leads to the development of arterial plaques.

This mechanism could be directly related to the creation of free radicals through the interaction of chlorine in the body. The free radical theory can also provide a reasonable explanation and link together the two major killers in the United States, heart disease and cancer. Free radicals are highly reactive molecular fragments generally harmful to the body.

In Super Nutrition for Healthy Hearts, Dr. Richard Passwater outlines the destructive nature of free radicals and how they are related to heart disease and cancer. The creation of plaque in the arteries and the creation of a cancerous tumor both have their origin in free radical activity. The plaque in atherosclerosis is essentially a benign tumor.

> The initial plaque formed by diseased arteries is due to a mutation of a cell in the artery wall. Certain chemicals in the bloodstream, including chemical pollutants, smoke components,

and reactive molecular fragments called free radicals, cause a normal smooth muscle cell in the arterial wall to go haywire (mutate). The fibrous plaque consequently formed is essentially a benign tumor. The origin of heart disease is akin to the origin of cancer.

The cell that mutated because of the presence of reactive chemicals reproduces itself (proliferates) in its new mutated form exactly, and all cells subsequently reproduced by this mutated parent cell are exact replicas (that is, they are monoclonal). They form a growth unique from normal arterial cells.

This growth is the first step in plaque formation that has for years been missed by other researchers. Dr. Earl P. Benditt of the University of Washington made this discovery using sophisticated techniques of cell differentiation with the electron microscope, coupled with chemical analyses of enzymes present in the cell.

Monoclonal proliferation continues at a rate faster than normal cell growth, and cell crowding occurs. A second stage is reached in which the crowded cells produce collagen and cholesterol to form a fibrous mass. This second stage is the uncomplicated plaque that for years has been wrongly considered the first stage.

The third stage is a complication of the second stage, in which the fibrous mass has erupted through the arterial wall into the bloodstream. In this stage, calcium and cholesterol are attracted from the bloodstream and added to the fibrous mass by electrostatic-charge attraction.

This complication adds to the plaque size and reduces blood flow, and also enhances blood clotting. (27)

From this description, we can see the link between heart disease and cancer through the medium of chlorine creating free radical activity in our body. Chlorine is one of many substances that can result in free radicals. Without discussing the many other substances, like excessive unsaturated fats, that promote this free radical activity, the harmful connection with chlorine seems self evident. If your drinking water is chlorinated, don't drink it. You can purchase very effective filtration devices which will remove 99% of the THM's or purchase water that chlorination has not been added to.

Just this simple safeguard - drinking only non-chlorinated water - may save thousands from heart disease and cancer - the two major degenerative killers in the United States.

7 • THE WATER STORY AND ANIMAL EXPERIMENTS

The research we have reviewed and analyzed has been from studies conducted on humans - epidemological, ecological and individual. They have led us to the conclusion: drinking water has a real affect on your health. There is a beneficial and/or protective factor present in hard water and water high in total dissolved solids and with an alkaline pH.

Because of the general nature of these human studies scientists have begun to aggressively use animals to verify and expand on the water story.

> Epidemologic studies have suggested factors in drinking water influence on the human cardiovascular system. A clear identification of the factors involved requires more invasive techniques and more strict experimental controls than can usually be applied in epidemologic studies. Consequently, laboratory animals are often used to expand and support epidemiological data. (1)

For the most part, the animal studies support the conclusions drawn from the human studies. However, like the human studies, animal studies are at times inconsistent. Usually, the inconsistency arises in comparing the effect and mechanism of the test substance between two different types of animals. For example, testing the effect of cadmium on rabbits, horses and rats may result in different effects and conclusions. We must keep in mind that the reaction of an animal to a test substance may be different than the human reaction. Yet, at the same time, animal studies are valuable because of their potential to explain and/or to pose theories that can be extrapolated to humans.

Many of the animal studies dealing with the water factor and heart disease have investigated the impact of cadmium (Cd) in hard water and soft water.

> Cadmium has been suspected as an etilogic agent for human hypertension. This element concentrates in the kidneys, and human hypertensives have been found to excrete more Cd in the urine than normotensives. High tissue Cd concentrations have been reported in human hypertensives at post morten examinations.(2)

Several studies have shown that rats given cadmium become hypertensive. "Elevated blood pressure without other evidence of cadmium toxicity has been induced in rats by long-term, low level cadmium feeding."(3) However, cadmium dissolved in <u>hard water</u>, rather than in distilled water, failed to raise blood pressure. In addition, the authors found that some minerals can act as a protection against the harmful effects of cadmium. Small amounts of selenium and large amounts of both zinc and copper inhibit the induction of hypertension by cadmium. Lead significantly augments rather than inhibits the response to cadmium. (4)

Most animal studies just look at hard water (water high in calcium and magnesium or CaCo3) vs soft water (water low in calcium and magnesium or CaCo3). They don't realize the potential synergistic effect of other minerals and total dissolved solids. Clearly, for rats, hard water and water with other minerals are beneficial in preventing cadmium induced hypertension.

The effect of cadmium and water on horses has also been closely observed. Fifty Swedish horses, slaughtered for meat production, were studied. The water tested was the actual horses' drinking water from their native habitat. The authors' studied cadmium concentrations in the kidney cortex and changes in the myocardium and aorta. Their findings:

> Cadmium concentrations in the kidney cortex were lower in horses drinking hard water than in horses drinking soft water, and a significant negative correlation was found between water hardness and Cd concentrations in kidney. (5)

Hard water is not only best for horses but it seems to be best for soil irrigation as well.

> . . . hardness of the drinking water reflects the pH of the soil . . . Soft water as a rule is more acid than hard water. In an acid environment, Cd from soil and fertilizers is more effectively absorbed by the plants. It is therefore possible that horses living in soft water areas are more exposed to Cd from locally grown food than horses in hard water areas. (6)

The implication of this observation should be carefully studied. The kind of water we irrigate our crops with could be very important to our health.

Due to the limited number of horses, the myocardium and aorta changes were not statistically significant. However, the authors state the following:

> . . . microscopic examination of the myocardium and aorta showed a greater frequency of changes in horses who drank soft water than in those who drank hard water, a finding which lends further support to the existence of a water factor associated with CVD . . . Microscopic

changes in the aorta and myocardium were approximately two times as frequent in horses that drank soft water compared to horses that drank hard water. (7)

In Borgman and Lightsey's study, rabbits were given water containing cadmium to determine the effects of both cadmium and hard water upon lipid metabolism.(8) The water used for one group of rabbits was de-ionized and distilled. For the other group, hard water was synthesized by adding 500 mg of calcium and 100 mg of magnesium per liter. Hair samples, taken at the beginning and end of the experiment, revealed an increase in the calcium concentration in rabbits given hard water, but no increase in cadmium. There was however, an accumulation of cadmium found in the liver and kidneys. Whether the water was hard or soft, the amount of cadmium accumulation was not influenced.

Why hair analysis did not reveal the increase cadmium is difficult to tell. The authors suggest it may be due to the length of the experiment. Twelve weeks may be too short a time to observe accumulations in the hair. However, it could show a weakness in using hair analysis as the only evaluative tool for toxic metals.

Perhaps rabbits metabolize cadmium in different ways and at different rates. How rabbits metabolize cadmium may indicate that hardness per se (Ca & Mg) is not enough. Other minerals interacting with calcium and/or magnesium may be needed to create the protective factor. However, there was one essential difference noted with regards to water. Rabbits given hard water resulted in lower liver cholesterol concentrations and less cholelithiasis (gall stones).

One of the most important studies was done on chickens and the effect of chlorine. This study is fully described in Coronaries/Cholesterol/Chlorine, by J. M. Price. Because of the ubiquitous presence of chlorine in our water and its relationship to both heart disease and cancer, a lengthy summary of Dr. Price's work will prove revealing. (9)

Two experiments were done in an attempt to prove that chlorine is the essential cause of atherosclerosis. Two groups of male chickens were used. Chickens have been widely accepted as an excellent animal for research in atherosclerosis.

Both groups ate cooked mash consisting of about 1:1 mixture corn and oat meals with 5% low-priced oleomargarine. Pure distilled water was used exclusively. Chlorine, in the form of chlorine bleach (about one-third teaspoonful per quart of water), was added to the drinking water and mash of one group.

Within three weeks, there were grossly observable effects in both appearance and behavior of the chlorinated group. The control group was in vigorous health. Some of the experimental group died after four months. Careful examination showed 95% of them to have grossly visible thick yellow plaques of atherosclerosis protruding into the lumens. After seven months, so few of the experimental group were alive, the rest were sacrificed. The findings were the same - 95% were grossly affected. At the same time, one third of the control group were sacrificed with not one abnormal aorta found.

Although these results seemed conclusive, it was decided to repeat the procedures by taking the healthy control animals of the first experiment and dividing them into an experimental group receiving chlorine and a control group. The outcome was identical!

In the chapter, Fluoridation and Cancer, reference was made to an experiment that studied the relationship of water hardness and fluoride toxicity in trout. Briefly, the study showed the softer the water, the more fluoride passes through the intestinal wall.

Using a variety of different waters, Neal and Neal did experiments on rabbits and atherosclerosis. In the first experiment, rabbits were fed a high-fat, high cholesterol diet. There were four groups of rabbits, each with the same diet but with different drinking water. Group 1 received distilled water; Group 2 received untreated hard sulfur water from artesian wells; Group 3 received hard chlorinated water from Jacksonville, Florida city water; and Group 4 received hard non-sulfur water from rock wells. After two months, the rabbits were sacrificed and atherosclerosis of the aortas evaluated.

> The aortas of the rabbits on distilled water had more atherosclerosis than those on hard water. It did not make any difference whether the water was hard sulfur, hard chlorinated, or non-sulfur hard water. The groups on hard water had less atherosclerosis than the group on distilled water. (10)

In the second experiment, the same high-fat, high cholesterol diet was used, but the water was different. Group 1 received distilled water; Group 2 received hard sulfur water; Group 3 received distilled water plus calcium carbonate; and Group 4 received distilled water plus magnesium sulfate. After three months, the rabbits were sacrificed and their hearts studied.

> Again we found there was more atherosclerosis on the average in the distilled water group (Group 1) than in the hard water group (Group 2). Group 3 rabbits (calcium carbonate) showed moderate atherosclerosis . . . while Group 4 rabbits (magnesium sulfate) showed no atherosclerosis. (11)

A similar study showing the protective aspect of magnesium on high-fat diets with rats was done by Vitale and his associates.

Magnesium oxide (MgO) given orally caused a reduction in atherosclerosis in rats on a high-fat diet. They also found that a high-fat diet caused magnesium deficiency in rats not given more magnesium than the usual requirement on a regular diet. (12)

An animal study by Ingols and Craft, evaluated the protective action of calcium and magnesium ions against the passage of metallic ions from food or water through the intestinal wall to the vascular system.

One group of mice was given hard water (300 mg/L), and another group drank soft water (30 mg/L) for three weeks. Manganese was present as part of the standard laboratory chow. The mice were sacrificed and their livers were examined for manganese by neutron activation. The manganese level in the livers of the mice drinking hard water was much lower than in the livers of the soft water mice. The feces from these soft water mice had a much lower manganese content than did the feces from the hard water mice.

This type of analysis was done for manganese, gallium and for a mixture for manganese, tungsten, copper and lead. In all three studies, the hard water mice group had less of the substance found in their livers and more was found in their feces, than the soft water mice.

The implications of this study are highly significant. Hard water seems to enable the body to metabolize the amount of minerals it needs and to excrete the balance. Even though so-called good minerals were excreted, which might be construed to eventually lead to mineral deficiences, "the better human experience with hard water . . . would imply that deficiences do not develop." (13)

Hard water appears to provide a protective mechanism against substances - beneficial or harmful. In this study, hard water prevented the accumulation of harmful substances; still allowing adequate absorption of useful minerals.

Recent research by Richard Bull, head of the EPA's toxicology and microbiology division in Cincinnati, points to the strong relationship between calcium, chlorine and heart disease.

> Preliminary animal research now strongly suggests that a diet somewhat low in calcium may be a risk factor in heart disease for that half of the U.S. population whose drinking water is disinfected with chlorine. (14)

A low calcium intake is very common. According to the HANES report (Health and Nutrition Examination Survey), most adults in the United States do not even meet the RDA for calcium.(15)

To test the calcium-chlorine relationship, Bull studied pigeons. Pigeons are a good laboratory model for the study of atherosclerosis in humans.

> Twelve pigeons were fed a diet that was normal except for its calcium content; it containd only 80 percent of the recommended daily allowance (RDA) of that mineral. Half these birds drank unchlorinated water, the rest drank water containing 10 milligrams of chlorine per liter. After only three months, serum cholesterol levels of the birds that drank chlorinated water were 50 percent higher - or 300 mg per deciliter of blood - than those of the birds that drank unchlorinated water. (16)

Birds who drank chlorinated water and took only 80 percent of the daily RDA of calcium, had a huge increase in their cholesterol levels.

A further test emphasized the protective quality of calcium on cholesterol levels, even if the birds drank the chlorinated water.

> . . . pigeons that ate a normal diet showed no statistically significant difference in cholesterol levels between birds with and without chlorine in their water; all had roughly 200 mg/dl. (17)

Calcium, one of the main ingredients in hard water, protected the pigeons from the harmful effects of chlorinated drinking water.

When dietary fat was added to the pigeon's calcium deficient diets, dramatic effects occured.

> . . . 10 percent lard was added to the calcium deficient diet. Though birds that drank unchlorinated water had only slight modified cholesterol levels - 230 mg/dl - pigeons that drank chlorinated water had 600 mg/dl. (18)

Hopefully, future research will look at the effect of fat, chlorinated water and calcium high water (hard water). Can the proper amount of calcium offset the negative effects of the fat (lard) diet?

The conclusion from these animal studies is very clear: hard water is healthier for you than soft water. Hard water, water high in calcium and magnesium, provides protection from a variety of potentially harmful agents such as cadmium, lead, chlorine and dietary fat.

8 • DE-MINERALIZED WATER: IS IT HEALTHY TO DRINK?

De-mineralized water is water that has had most, if not all, the minerals removed through the process of distillation, reverse osmosis or ion-exhange. De-mineralized water is soft water because it lacks minerals.

Many authors who have written nutrition oriented books on drinking water have proclaimed minerals in drinking water are very harmful to your health and should be removed. These popular books include such titles as:

The Shocking Truth About Water,
by Patricia and Paul Bragg
Water for the Eighties: A Cause for Concern,
by Eldon C. Muehling
The Choice is Clear,
By Allen E. Banik
The Great Water Controversy,
by T. C. Fry, Herbert M. Shelton, et al.
Water Can Undermine Your Health,
by Norman Walker
You Have a Right to Know,
by Esther Dougherty

All of these books have several features in common. They all strongly advocate de-minerlized water as the only water to drink. They all seem to be unaware, or else ignore, the research on drinking water and its relationship to heart disease and cancer that has gone on for more than twenty years. They all base their belief on the notion that the body cannot metabolize inorganic minerals and therefore, these minerals become deposited in the arteries and joints of the body leading to heart disease, arthritis, kidney stones and eventually death. They all seem to believe that even if the minerals in the water could be utilized, they constitute such a small percent of our requirements that its not significant.

Because of these books, many people have been led to believe that drinking de-mineralized, soft water is the only safe and beneficial water to drink.

The research presented here on heart disease and cancer clearly does not support the drinking of de-mineralized, soft water - to put it mildly! In addition, the animal studies demonstrate the potential harmful effects of soft water.

There is no question that polluted water is harmful and dangerous to our health. Pollution can come from chlorination (THM's), fluoridation (sodium fluoride), lead, cadmium, mercury, synthetic organic compounds, pesticides and so on. Yet, there are ways of getting good drinking water - either by cleaning your water with proper filtration systems or by purchasing bottled water. The next chapter, Which Water is Best to Drink?, will outline what to look for in bottled water and filtration systems.

Recently, the Organic Consumer Report, written by Dr. Betty Lee Morales and her husband, John Clark, discussed the drinking water of the Vilcabambas tribe and the Hunzas. Both of these people are known for their long, vibrantly healthy lives. Both drink highly mineralized water!

> What appears to be the chief reason for good health, and mineral balance is attributed to their mineral-rich waters, and especially their content of the very minerals that are most missing, or dangerously low, in most commercial

and public waters. (1)

The minerals they listed are: manganese, chromium, selenium, zinc, potassium, magnesium, phosphorus and calcium.

> The Vilcabambas as do the Hunzas, drink mineral rich water from mountain streams which flow underground and on the surface, over rocky formations which release not only calcium, but all other minerals, and trace-elements. Electrically charged, positively or negatively, from the friction of flowing and tumbling waters unpolluted, they provide assimilable minerals.(2)

Those authors who advocate de-mineralized water, like to emphasize that even if the minerals were useful, drinking water constitutes only a minor contribution to our daily intake of essential minerals and trace elements. They state that food sources contain all our mineral needs and therefore, minerals from drinking water are insignificant. But, there are growing indications that the minerals in water may be more bio-available than minerals in food!

> . . . growing attention is being paid to the chemical form in which trace elements are present and to the influence of these forms on biological availability and activity. The essential trace element chromium, for example, can occur in water in the hexavalent state and, as such, would be better absorbed by the organism than the trivalent form present in food. (3)

An interesting and thought provoking study on mineral absorption was done by Dauncey and Widdowson. They evaluated the absorption of calcium, magnesium, sodium and potassium in Englishmen living in hard and soft water areas. Urinary excretions were used as a measure to approximate mineral absorption. The authors' findings:

> The amounts of calcium, magnesium and potassium excreted in the hard and soft water areas were not significantly different. The excretion of sodium was significantly higher in the soft water areas . . . This was almost certainly due to salt added to the food. (4)

The greatest difference in drinking water was in calcium. The London men took in 110 mg of calcium per day, compared with the 3 mg of calcium for the Glasgow men. How do we reconcile the authors' findings above with the fact that the London men took in greater amounts of calcium in their water and diet than the Glasgow men? And with the authors' conclusion "that London men absorbed no more calcium than those in soft-water areas." (5)

One would expect to find a greater amount of urinary excretion of calcium from the men who took in more calcium. Yet, this is not the case.

Looking at actual figures, we find that the London men excreted 236 ± 8.2 mg and the men from Glasgow excreted 271 ± 14.7 mg of calcium. The London men took in more calcium from water and diet than the Glasgow men, yet excreted less calcium. Surely, the indication is the men from London absorbed more calcium than the Glasgow men. And, more importantly, this supports the research on the value of calcium in drinking water.

Crawford found a decrease in cardiovascular disease in England in areas with hard water which was best correlated with the amount of calcium in the water.(6)

We can no longer believe minerals in drinking water are unimportant. We have seen their role in heart disease and cancer and the picture is broadening. The minerals in drinking water may provide an important difference in many people's diets. Dr. Masironi, from the World Health Organization (WHO), states:

> Usually modern diets are marginally deficient in many elements, and therefore, the contribution of hard water or a daily intake of lithium or magnesium, or particular other elements may make a difference between optimal and sub-optimal intake. Also, chemical forms of many of these elements in water is different from what is in food. In food, there may be some chelating agents that prevent immediate absorption of certain elements and even if they are contained in very high amounts in food, their absorption comes down to perhaps a few percent. . . A study was done in India, where the food is deficient in calcium, and these people get most of their calcium from the hard water they drink.(7)

The research papers discussed here have shown the relationship of drinking water to heart disease and cancer and the importance of a variety of minerals for good health. Some of the minerals that have been highlighted to have beneficial effects are: silica, magnesium, chromium, lithium, vanadium, calcium and zinc. (8-15) The importance of a mineral in one study may be different than in another study. This may be due to the inherent geographic and social factors of the people in each area. Yet, it seems well founded that minerals in drinking water play a very important role in our health.

Mineral absorption is greater from drinking water than from food according to Dr. John Sorenson, Medical

Chemist and Professor at the University of Arkansas whose specialty is the metabolism and pharmacology of inorganic minerals. "Minerals in drinking water are better absorbed. For gastric intestinal absorption, things in solution are well absorbed. Simply because water is well absorbed, anything in it would also be." (16)

The metabolism of essential vs. non-essential metalic elements (zinc, lead, cadmium, copper and so forth) is greatly affected by the presence and amount of the essential element. If the needed essential element is present, there is less incorporation of the non-essential element. Therefore, more of the non-essential element is excreted. This is exactly the pattern we observed in the effect of different types of waters on animals. For example, if you have sufficient calcium, magnesium and other essential elements present in the drinking water, along with lead and cadmium, there is a higher probability that the essential elements will be absorbed and not the non-essential. In this way, the essential elements can provide a protection against the absorption of non-essential, and potentially harmful elements.

Further, if lead and zinc are both present in the drinking water, zinc would be selected by the cell and protection against lead would be provided. But, in the event there is no zinc, then as a second choice, lead could be selected, which would result in a dysfunctional protein or enzyme. If the protein or enzyme being synthesized by the cells requires a metal element and if the require metal is not there, or is there but in insufficient amounts, the non-essential metal can end up in the protein or enzyme. And, if this occurs, the protein or enzyme would be dysfunctional and toxic. (17)

The metabolism of metalic elements reveals the importance of drinking water with a high degree of total dissolved solids and hardness. This metabolic understanding reinforces the epidemological research dicussed previously.

Another procedure the body uses to absorb inorganic minerals is by "natural chelation, the suspending of minerals in hydrolized protein." Dr. DeWayne Ashmead, in Chelated Mineral Nutrition, outlines the steps.

(1) The mineral is ionized or broken apart from its carrier by going into solution. Example: calcium lactate is split into basic calcium, carbon, hydrogen and oxygen.

(2) The protein, which must be present, is broken down to amino acids by the hydrolyzing effect of hydrochloric acid and other digestants.

(3) With proper changes in pH, free amino acids attach temselves around the bare mineral ion and form an easily absorbed molecule. (18)

For these steps to take place in the stomach and/or intestines, many factors must be present. Dr. Ashmead continues:

(A) Minerals and proteins must be ingested at the same time.

(B) Adequate digestive aids must be present to ionize the mineral and also break down the proteins to amino acids.

(C) The amino acids must combine in a stable formation with the base mineral, that is, suspending the mineral ion between two or more amino acids from hydrolized protein. This necessitates an abundance of certain amino acids.

(D) The factors which frequently interfere with this chelation process must be eliminated or at least controlled. (19)

So minerals must be chelated in the stomach and/or the intestines with amino acids from digested or hydrolyzed protein before they can be absorbed.

Many people, especially as they grow older, do not produce adequate hydrochloric acid in their stomach and also have poor protein intake. These factors could result in very poor mineral absorption and lead to a variety of degenertive problems. The problem in not with the minerals in the drinking water but with the nutritional and metabolic deficiencies.

For example, some authors, who advocate de-mineralized water, suggest it would help prevent or eliminate kidney stones. However, studies done on the relationship of kidney stones and drinking water have not found this to be so. In fact, the exact opposite is true - hard water, not soft water is more beneficial.

Kidney stones diagnosis was significantly increased in soft water areas, especially where water hardness was less than 50 ppm. (20, 21)

So far, we have been looking at the reasons for drinking de-mineralized water and have found the reasons advocated for its use are not true. An additional problem with de-mineralized water can come from cooking your food in it.

The type of water you cook your food in will effect the calcium level in the food. Studies have been done on potatoes, cauliflower, carrot, endive and snap beans. (22,23) The vegetables cooked in hard water had significantly higher levels of calcium than those vegetables cooked in soft water. People who boil vegetables in soft water and throw out the water are losing most of the calcium from the vegetables.

The cooking factor may contribute and amplify the earlier observed findings on heart disease and hard water. Namely, heart disease is higher in softer water areas than in hard water areas. Cooking with soft water and having a low calcium intake may be responsible for over 10% of the U.S. population not meeting the minimum RDA requirements for calcium. Drinking water may contribute up to 20% of a person's daily calcium intake.

> Drinking water may contribute 10 to 20% of the total daily intake of calcium. And, recent studies by the USDA's Agricultural Research Service found calcium intake was 12% below the recommended amount. (24)

One writer, Linda Clark, who recognizes the harmful effects of drinking de-mineralized water, recommends adding a portion of sea water to distilled water to restore its mineral balance. (25) Certainly, this recommendation would be better than drinking distilled water alone, but whether this would restore the water to its proper mineral balance is questionable. If you decide to do this, you would be wise to verify what type of water you have created and follow the guidelines in the next chapter on total dissolved solids, hardness and pH. It is highly doubtful that you can create a water that is palpable by adding sea water in the amount needed to create water sufficiently high in total dissolved solids and hardness.

There appears to be no rational reasons nor scientific studies to support the beneficial use of drinking de-mineralized water. Before drinking de-mineralized, soft water carefully consider the numerous animal experiments (see Chapter 7) and the information presented here.

Be aware! For very different reasons, polluted water and de-mineralized water can be harmful to your health. Which water is best to drink is the subject of the next chapter.

9 • WHICH WATER IS BEST TO DRINK? AN EVALUATION OF BOTTLED WATER AND HOME FILTRATION SYSTEMS

Our discussion of the water story and its relationship to heart disease and cancer leads us to the obvious practical question, "Which water is best to drink?"

There are many choices: tap water, filtered water, distilled and reverse osmosis water, bottled water, bottled spring water and mineral water. Mainly, because of the chlorine added to municipal city water systems, unfiltered tap water can be dangerous to your health.

If you have your own well, it is wise to have your water checked regularly for a variety of toxins and carcinogenic agents. Compare the analysis of your drinking well to the items listed in the FDA Bottled Water Standards (Table 8) and to the National Interim Primary Drinking Water Standards (Table 9). These tables appear in the later part of this chapter. By using both standards, you will have an excellent measuring rod to evaluate the quality of your water. If your water exceeds these standards, start asking questions and determine if the excesses are a cause for alarm.

Many people have heard the warnings of the dangers of tap water. Many decided to purchase bottled water. Most of these people buying bottled water find the taste better than tap water. However, taste, by itself, is not a valid criteria to determine if a type of water is beneficial or harmful for you. The taste factor certainly should be used, but only after you have properly evaluated the water.

Our research has located three main characteristics in drinking water associated with health benefits.

They are:

(1) Total dissolved solids (around 300 ppm)

(2) Hardness (around 170 mg/L of calcium carbonate)

(3) Alkaline pH (over 7.0)

(Note: ppm (parts per million) and mg/L (milligrams per liter) are the same amount and can be used interchangeably.)

Does bottled water fulfill these requirements? To understand this, let's briefly look at the bottled water industry as a whole and review the various types of bottled water available. Not all bottled water is the same!

Water today looks like the oil business in the mid-1950's.(1) The sales of bottled water have increased 93% during the last five years. In 1981, gross sales were $700 million with fifty-five million gallons of water sold. Twenty-seven million gallons of water were imported, with Perrier capturing three-fourths of the market. One out of eighteen families in the U.S. buys bottled water, and, in pace-setting California, one out of three. Of the total U.S. sales, Florida, Illinois, California, New York and Texas account for 87%, with California responsible for over half.(2) All of this activity has made water the fastest growing beverage on the market.

Water is big business and big business owns some of the major bottled water companies. For example, the top three bottled water companies in the USA are all owned by corporate giants: Coca Cola Bottling Company of Los Angeles recently sold Arrowhead Puritas Company to Beatrice Foods Company; Sparkletts Drinking Water Corporation is owned by Foremost-McKesson; and Great Bear Spring Company is owned by Coca Cola Bottling Company of Los Angeles. According to Business Week, Coke does more than $100 million a year in the in the water business.

Bottled water is a huge industry. With heavy advertising, the consumer can sometimes be misled as to the true value and quality of bottled water. Consider this:

> More than 700 brands of bottled water are now available in the United States. About 80 percent of these are not 'natural' but are treated or processed water, usually filtered or distilled well water.(3)

Since July 1, 1979, bottled water sold in interstate commerce is under the jurisdiction of the Food and Drug Adminstration (FDA). The FDA chemical standards for bottled water are:

TABLE 8

FDA BOTTLED WATER STANDARDS (4)

Substance-Concentration in Milligrams per Liter

Substance	Concentration
Arsenic	0.05
Barium	1.0
Cadmium	0.01
Chloride	250.0
Chromium	0.05
Copper	1.0
Iron	0.3
Lead	0.05
Manganese	0.05
Mercury	0.002
Nitrate (N)	10.0
Phenols	0.001
Selenium	0.01
Silver	0.05

TABLE 8 (Continued)

Sulfate	250.0
Total Dissolved Solids	500.0
Zinc	5.0
Organics	
Endrin (1,2,3,4,10,10-hexachloro-6,7,-epoxy-1,4,4a,5,6,7,8,8a-Octa-hydro-1,4,-endo, endo-5,8-aimethano naphthalene	0.0002
Lindane (1,2,3,4,5,6,-nexachloro-cyclohexane gamma isomer)	0.004
Methaoxychlor (1,1,1-trichloro-2,2-bis [p-methoxy-phenyl] ethane)	0.1
Total Trihalomethanes	0.10
Toxaphene (C10H10CL8)-technical chlorinated camphene, 67-69 percent chlorine)	0.005
2,4-D (2,4-dichlorophenoxyacetic acid)	0.1
2,4,5-TP Silvex (2,4,5-trichloropheno-xypropionic acid)	0.01

Bottled water that exceeds these standards must be so labeled. See Appendix Seven for the complete standards which include fluoride levels, microbiological, physical, and radiological specifications.

In the early 1970's, an EPA survey of twenty-five USA bottled water plants showed:

(1) Sanitary deficiencies in all facilities surveyed.
(2) Bacteriological surveillance was judged inadequate

in almost every case.

(3) 8% of the samples had coliform organisms.(5)

Since then, the FDA has established quality standards for bottled water which are essentially equivalent to the EPA standards for tap water. The compliance procedures for bottled drinking water state that each plant is to test for bacteria at least once a week and perform chemical, physical and radiological tests annually.(6) In discussion with the FDA and the International Bottled Water Association (IBWA), their agents state that the FDA inspects bottled water plants annually.(7) In addition, members of the IBWA are inspected annually by the American Sanitation Institute.(8) The IBWA members represent 85% of the total bottled water sales in the USA. Besides domestic bottled water plants, members include foreign bottlers who sell and distribute imported water in the US.

Although there has never been a follow-up report by the EPA or FDA on the bacteria problem uncovered in the EPA 1972 survey, the current standards and procedures appear very adequate in insuring that bottled water is free from harmful bacteria when bottled. According to the EPA, "there is no monitoring program from bottled water as there is for tap water . . ."(9) However, this should not be a cause of any alarm. The standards and procedures set up by the FDA are similar to the way all food and beverage inspection is handled. In short, the so-called bacteria problem in bottled water today is not geniune.

There are various <u>types</u> of bottled waters. The following definitions for tap water, processed water and natural spring and mineral water are useful and informative.(10)

I. TAP WATER

- may originate in a lake, river, reservoir, or underground well.

- is treated with chemicals, such as chlorine, to combat environment pollution, unless its your own private well.

- often contains industrial contaminants and organic solids.

- may be adversely affected by the quality of the pipes through which it flows.

All municipal water systems must meet federal state and local water quality standards. However, tap water quality and taste vary from community to community, and from tap to tap depending on the source, type of processing and quality of plumbing.

II. PROCESSED WATER

- is usually tap water which has been purified or treated to remove chemicals, sediment and minerals.

- may originate in a stream, well or municipal tap. The source is usually not indicated on the label.

- is treated by filtration, reverse osmosis, chemical purification, distillation or some other method.

- may be non-carbonated, such as the bulk water found in many office water coolers; or may be carbonated, such as seltzer or club soda.

Processed waters vary in quality, taste and purity, depending on the source of the water, type of treatment, and bottling processes. Distilled water, is processed by evaporation and condensation.

Some better known brands of processed water include:

- Canada Dry Club Soda
- Schweppes Club Soda
- Ozarka Drinking Water
- Sparkletts Crystal-Fresh Drinking Water
- Deep Rock Artesian Fresh Drinking Water

III. NATURAL SPRING WATER AND MINERAL WATER (May be still or sparkling)

As known in the United States, natural spring water

- is unprocessed water drawn from one of several springs or other natural sources.
- may or may not be bottled directly at its source.
- may be bottled in containers of any size.

Some better known brands include:

- Artesia
- Saratoga
- Poland Springs
- Mountain Valley
- Deer Park

In California, for a water to be labeled 'mineral water', it must contain 500 ppm of total dissolved solids. This can either be a processed water or a natural water. The FDA has not officially defined mineral water.

To be called mineral water in Europe, a bottled water must:

- flow freely from its source. It may not by pumped or forced from the ground.

- be bottled from one souce only, which must be listed on the label.

- be bottled directly at its source, in containers with a maximum size of 2 liters. It may not be stored prior to bottling.

- show proof that its temperature, mineral balance and water pressure have not varied in ten years.

- be constantly monitored by the ministry of health.

Some better known brands include:

- Evian (France)
- Perrier (France)
- Ferarrelle (Italy)
- Apollinaris (West Germany)

Now that we understand the types and definitions of water, how do we evaluate which of these water is best for our health? Based on the information presented on water and its relationship to heart disease and cancer, and assuming you want to purchase bottled water, follow this checklist for water selection.

BOTTLED WATER CHECKLIST

(1) Obtain a complete water analysis from the company.

(2) Is the lab report from an independent lab or simply an in-house lab report done by the company itself?

(3) Ask how frequently these tests are performed.

(4) Check to see if the water is processed or natural. If its processed, don't buy it. If its natural, evaluate further.

Processed water can be filtered water or de-mineralized using distillation, reverse osmosis or ion-exchange. In either case, you can do this for yourself and save substantial money. Some filtration systems are excellent and will be discussed shortly. (In the chapter, <u>De-mineralized Water: Is It Healthy To Drink?</u>, the negative and unhealthy aspects of de-mineralized, soft water are thoroughly discussed.)

Most processed water companies first remove the total dissolved solids in the water and then add back a selection of mineral salts. This is like refined flour - removing twenty or more vitamins and minerals and then adding back five or six and calling it enriched. Remove all the dissolved solids from the water and add back five or six mineral salts and call it bottled, spring-like or crystal-fresh water, or some other mis-leading name. Don't be fooled; read the reports and if its processed water, don't buy it!

(5) Check to see if any independent agency has published

an analysis of the water. For the reader's convenience, three excellent and hard-to-find reports are in Appendix 1, 2 and 3.

(6) Once you have collected this data, you will have an excellent profile on the water. Now, record the following characteristics and see how it compares to these recommendations:

(A) Total Dissolved Solids (TDS) - around 300 ppm.

(B) Specific Conductivity - around 500 micromhos or higher (65% of the specific conductivity will give you the total dissolved solids. This can be useful when you don't know the total dissolved solids but the specific conductivity is given.)

(C) Hardness - Usually if total dissolved solids is high, the water will be hard. But, check the lab reports, noting the calcium and magnesium or CaCo3 (calcium carbonate) levels. Four categories of hardness, expressed as CaCo3, in common use are:

0 - 60 ppm - soft
61 - 120 ppm - moderately hard
121 - 180 ppm - hard
181 or more - very hard

Based on recent reports from Great Britain, 170 ppm of hardness was found to provide the best health benefits.

(D) pH - alkaline, above 7.0

(7) Once you have done this, you will want to check the lab reports for specific toxins and carcinogenic agents. Some items to check for are: THMs, artificial fluoridation (sodium fluoride), asbestos, lead, cadmium, mercury, arsenic, and synthetic organic chemicals, such as: PCBs, carbon tetrachloride, pesticides and nitrates.

Check the FDA Bottled Water Standards and compare it to the analysis of the water your are considering. Top quality bottled water should not contain any THMs, synthetic organic chemicals or pesticides. Just because its bottled water, doesn't automatically mean its devoid of harmful substances. For example, in 1982, Spectrix Corporation, an EPA-certified laboratory from Texas, tested five water samples - Evian, Perrier, Artesia (three natural spring waters); Houston tap water - ground source; Houston tap water - surface source; and Ozarka (a processed water). The results were Evian, Perrier, Artesia and Houston tap water - ground source, all showed no THMs. Houston tap - surface source, had 13 ppb and Ozarka had 28 ppb of THMs. (11) The EPA suggested limit is 100 ppb (parts per billion).

(8) Now that you have narrowed down your list of acceptable waters, its time to taste them.

Each of us have different preferences and they can vary widely. Go by your own taste. Please remember that harmful chemicals in water are not necessarily apparent to our taste.

Go through the steps outlined and you can have confidence that the water you have selected is helping you maintain and gain better health and is not going to result in potential health problems.

An alternative to buying bottled water is using your tap water and a filter system. Evaluating home filtration systems can be as complex as evaluating bottled water. Household water treatment units are over a $1 billion a year business.(12) And, as you can imagine, there are hundreds of units available, each claiming to do wondrous things.

Reading the manufacturers' claims and studying their lab reports can be a frustrating experience. Each product comes with a different style lab report; some reports evaluate some items but not others. In short, its a confusing mess! How does an intelligent, concerned person evaluate the claims of these products?

Some of the best overall guidelines and advice in understanding water lab reports comes from Richard T. Williams. The following points are basically from him and, in some cases, have been slightly changed or shortened.

WATER FILTERS/PURIFIERS: WHAT TO LOOK FOR IN A LAB REPORT(13)

(1) Chemical removal tests should be conducted for the full rated life of the cartridge. For example, if a cartridge is rated for 500 gallons, the test should be conducted for 500 gallons.

(2) Lab reports should be checked to see what the minimum detection level sensitivity their equipment can measure. For example, a minimum detection level of 5 ppb (parts per billion) is meaningless when the maximum acceptable level is 2, 3, or 4 ppb.

(3) Input test levels are frequently so low, absolutely, or relative to detection sensitivity levels, that the

tests are meaningless. For example, a test input level of 400 ppb with a test sensivity of 200 ppb can be reported as 'none detected' with as much as 199 ppb of the substance still in the water. 'None detected', to the layman, implies complete removal, while in reality as much as 50% of the contaminent could remain in the water.

Also, a test input level of 20 ppb can be misleading if the contamination levels in a given area are 200 or 400 ppb or some other larger concentrations.

(4) Laboratory tests generally use specially prepared water rather than the tap water. While the use of special water is frequently necessary for research purposes, it can pose some problems and give misleading information. The chemistry of water is complex and ions and compounds frequently interfer with, or mask, other compounds, causing different results. For best results, water should be tested both ways: with special polluted waters and actual tap waters.

(5) Tests should note the removal capability of the product at the beginning, middle and end of the rated cartridge life for each compound or ion tested.

Using these guidelines, you should request product lab reports and see how closely they have followed these points. In some ways, asking for a lab report is like asking for a letter of recommendation for a job. No company sends you a poor lab report; just like no one gives a job reference from someone who may give them a poor recommendation. Because of this inherent bias, and because each company has its product tested to their own specifications, it is difficult to accurately compare one product to another.

Because of this confusion, two EPA sponsored research reports need to be highlighted. Anyone considering buying a filtration unit should be aware of this research.

In July of 1980, the EPA awarded a contract for $540,000 to Gulf South Research Institute to evaluate 31 home water treatment units. In Appendix 4 and 5, you will find sections of these reports. Although there has been some criticism directed at the findings, it does provide one of the best sources of reliable information available. All the products were tested the same way, so you can evaluate and compare the results.

Today many products are on the market that were not included in this report. Also, many improvements have been made on the tested filters. Read and check the overall performance of the units tested and you'll have a solid basis to evaluate all filters.

From these reports, it is clear that some products do an excellent job - up to 99% removal of THMs and synthetic organic chemicals - while others are clearly a waste of money. In fact, some products are so poor, they give the buyer a false sense of security.

An alphabetical selection of some of the better filters follows: (14)

	PHASE 1 (see Appendix 4)		PHASE 3 (see Appendix 5)
	% THM Reduction	% NPTOC Reduction	% Halogenated Organic Reduction
Aquacell	86	23	97
Aqualux CB-2	98	23	99
Continental 350	99	87	99
Culligan SG-2	89	28	99
Everpure QC4-THM	99	55	99
Seagull *	70	30	97

* Seagull was tested for 1600 gallons in Phase 1. Company literature claims the unit is for 1000 gallons. In Phase 3 it was tested for 1000 gallons. Therefore, the THM and NPTOC reductions would have been greatly increased had the unit been tested in Phase 1 for 1000 gallons as it should have been.

You will notice, in the above listing, a category for the reduction of NPTOC (non-purgeable total organic carbon). NPTOC is used as a background indicator for the total organics in the water.

> NPTOC describes the bulk of organic material found in water. Since these constituents are non-purgeable (exhibit low volatility), they can be generally classified as relatively high molecular weight species . . .
>
> The heavier organic chemicals evaporate less easily and are of unknown or poorly documented toxicities or long-term health hazard, whereas the organic chemicals of high volatility are lighter . . . and have more well-documented toxicities. (15)

In selecting your filtration system, you may want to give some attention to the filter's ability to remove NPTOC. Presently, there doesn't appear to be any health risks associated with NPTOC in drinking water, but who knows what future research will reveal. Certainly, the most important reduction is THM, but you may gain added safety by selecting a unit that removes high percentages of both THMs and NPTOC.

A selection of the lighter and more volatile organic chemicals were analyzed in the Gulf South Research Report. Different chemicals were analyzed in ground water and in surface water. The chemicals analyzed in ground water were: 1,1,1-trichloroethane, carbon tetrachloride, trichloroethylene and tetrachlorethylene. And, those chemicals analyzed in surface water were: p-dichlorobenzene, hexachlorobenzene and chlordane. For the most part, the filter's effectiveness in removing these organic chemicals were similar to the results on THM removal. However, because of the variety of chemicals tested, the reader is asked to look at Appendix 5 for the complete results.

By using the findings from these two groups of tests - THM and organic chemical removal - you will be able to select a water filter that does what you want it to do. You will know what your paying for and what it can and cannot do!

The build-up of bacteria on carbon filters and subsequent potential health risks has been a topic of concern also. In the Gulf South Research Report, tests were conducted for SPC bacteria (standard plate count) and endotoxins.

> SPC bacteria are normally present in tap water and are considered to be non-pathogenic (non-disease causing) under normal home usage situations. However, they are known to multiply on activated carbon and other surfaces and in water pipes in the absence of active disinfectants . . . Average increases in SPC populations . . . were around one order of magnitude and are not known to have any health significance . . . Silver-impregnated activated carbon units . . . do not significantly reduce bacterial growth patterns. (17)

Bacteria does multiply on carbon filters but the amount of bacteria created appears not to have any health significance. Silver-impregnated carbon filters do not reduce bacterial growth.

Besides bacteria growth, endotoxins are present in drinking water. They do not, however, appear to pose any health problems.

> SPC bacteria in drinking water release endotoxins from their cell walls upon their death . . . endotoxins are presently not known to be a hazard in drinking water . . . (18)

No significant elevation of endotoxins were found from using the filters tested. Again, silver-impregnated filters showed little, if any, differences in the endotoxin levels compared to non-silver filters.

The tests results for SPC bacteria and endotoxins were confirmed in the follow-up study in Phase Three of the Gulf South Research Report. These findings, however, point out the value of flushing and not using the first water coming out of the filter units.

> . . . the highest bacterial populations were found in the first water through the filter and justified the use of the first flush water in the basic laboratory testing program. (19)

Many people may be confused about the meaning of 'water purifiers'. Technically, a product labeled as a water purifier must be able to completely remove, kill or de-activate "all primary pathogens - bacteria, viruses, protozon cysts, worm eggs, etc. - everything capable of causing infection and disease in healthy people . . . A water purifier is a device which reliabiy makes microbiologically unsafe water, safe."(20) Some manufacturers and sellers of filter units claim theirs is a true 'water purifier'. This is probably not so. According to one industry report as of September, 1983, no simple point-of-use device is recognized by the governments of both the United States and Canada as a 'water purifier'." (21)

By flushing your filter daily and changing the cartridge regularly, you can protect yourself from any potential bacteria build-up and health problems. Each filter cartridge is rated for a specific number of gallons. Calculate the amount of water run through the filter per day to know when you'll need to change the cartridge. In general, one gallon per adult per day is a safe figure, assuming drinking and cooking uses. Don't wait till your filter clogs or the water flow slows down before changing the cartridge.

The following checklist from Consumer Reports is very useful and practical and will help you gain the benefits of filtered water without the risks.

(1) Flush out the filter before the first use of the day. Open the faucet wide and let the water run at least 30 seconds for an under-sink model, at least 10 seconds for a sink-mounted filter. When you install a new cartridge, flush for several minutes to remove fine carbon particles.

(2) Change filters regularly. A heavily used filter is more likely to contain high bacterial levels and to discharge organic chemicals previously trapped. An exhausted filter is worse than useless.

(3) Don't filter hot water. A filter on the faucet that passes hot water won't remove contaminants very well. And the hot water may liberate chemicals previously trapped on the filter . . .

(4) After installing a new cartridge, circle on your calendar the date for the next replacement. Stick to your schedule."(22)

The Gulf South Research Report did not evaluate the effectiveness of filter units in removing specific inorganic minerals or dissolved solids. Its too bad they didn't. Some manufacturers claim that their filter can remove the potentially harmful elements, like lead and cadmium, yet not remove the beneficial elements, such as calcium and magnesium. This is an area of confusion and needs to be straightened out.

A carbon filter cannot significantly reduce any total dissolved solids. If lead or cadmium or any of the heavy metals are removed by these filters, it is because these metals have become attached to particulate matter in the water and filters can remove particulate matter. Also, lead and other heavy metals have a tendency to attach themselves to the inside container of filter units.(23) This explanation may very well explain why some lab reports show lead and heavy metal reductions. Some filter units are using electrically charged materials (positive and negative charged fibers) to attract and remove undesirable pollutants. The effectiveness of this process appears highly useful and should be more widely studied.

The comparison of filter units in the removal of lead, cadmium, mercury and so on, is very difficult. You can receive specific individual reports from different filter manufacturers showing significant percentages of heavy metal removal.

The consumer has no true, reliable way of comparing one unit against another in this area. In the Gulf South Research Report, you could compare each filter unit's capability of THM and organic compound removal, because each unit was tested exactly alike. Such tests have not been done on filter units for heavy metals or inorganic elements. Technically, what is needed is testing of actual tap water and spiked water before and after filter use for all the standard inorganic elements - e.g. calcium, magnesium, copper, lead, cadmium, and so on. This type of testing could be expensive but is necessary if we are to truly compare and evaluate the full capabilities of filter units to provide us with healthy water.

By following these procedures, we would be able to compare every filter unit. We would also know if the units claiming high reductions in the heavy metals through electrical attraction affect other inorganic elements as well.

It may very well be that the units that performed the best with regards to THM removal and organic compounds, will also perform the best regarding heavy metal removal. But, ideally, these tests need to be done; either by industry or government or through joint research. With these tests, consumers should be in a much better position to evaluate which filter is best for their own purposes.

Another question of potential concern is, "can the amount of total dissolved solids affect the performance of filters?" Waters with low TDS have a tendency to pull matter off carbon filters. This can pose a problem to consumers who use portable filter units. For example, you are using a portable, carbon filter and your drinking water is high in TDS. You go on a camping trip, taking your filter with you, and the drinking water here is low in TDS. The water low in TDS can cause leaching, where the accumulated organics and heavy metals on the carbon end up in your drinking water! (24)

Even with these cautionary remarks, we still have adequate and reliable information from the Gulf South Research Report and others to safely recommend home filtration units.

But, before buying a filter unit, first check the quality of your present water. Remember, you are looking for the positive health promoting qualities in water - not just the removal of negative, harmful substances. Do you recall the animal studies discussed in Chapter 7? Animals given harmful doses of lead or cadmium were protected from the negative effects of these elements if they drank hard water. For most of us, lead and cadmium intake is mainly from food and/or air, not from drinking water. Certainly, we need to be careful in this area. For most homes, standard lab tests from your municipal water plant may prove adequate. However, it is safer and more accurate to have your tap water tested. Most state health agencies will do this for a fee. If you would like your tap water tested, please write the author for a list of reliable labs. (Author's address is in the Postscript.)

The research presented has offered these guidelines for total dissolved solids (300 ppm), hardness (170 ppm), and an alkaline pH (over 7.0). If your present water meets or is close to these parameters, consider using a filtration unit to remove THMs and organic chemicals. If you present water does not meet these criteria, buy bottled water that does.

You can receive a complete water analysis of your water from your city water department. If you have your own well, have it analyzed. With this data, you'll be able to see if your water meets the criteria. The city water report will list a variety of chemicals. The levels of arsenic, barium, cadmium, lead, mercury, nitrate, fluoride, THMs, organic compounds and pesticides should be well below the so-called safe level standards in the National Interim Primary Drinking Water Regulations. A complete copy of these standards can be found in Appendix 6. The maximum contaminant levels (MCL) for inorganics and organics are:

TABLE 9

NATIONAL INTERIM PRIMARY DRINKING WATER STANDARDS (25)

INORGANICS

CONTAMINANT	LEVEL (mg/L or ppm)
Arsenic	0.05
Barium	1.00
Cadmium	0.010
Chromium	0.05
Fluoride	(see appendix)
Lead	0.05
Mercury	0.002
Nitrate (as N)	10.00
Selenium	0.01
Silver	0.05

All of the above MCL's, except nitrate, apply only to community water systems. The nitrate level applies to both community and non-community water systems.

ORGANICS

Endrin	0.0002
Lindane	0.004
Methoxychlor	0.1
Toxaphene	0.005
2,4-D	0.1
2,4,5-TP Silvex	0.01
TTHM	0.10

All of the above apply only to community surface water systems.

Based on the information presented throughout this report, these standards should be viewed with a high degree of caution. There are no methods to establish a threshold for long-term effects of toxic substances. As the National Academy of Sciences' report on Drinking Water and Health, warns:

> The insidious effect of chronic exposure to low doses of toxic agents are difficult to recognize, because there are few, if any, early warning signs and, when signs are ultimately observed, they often imply irreversible effects. (26)

Standards have a way of changing due to political pressure and money. Don't let these standards give you a false sense of security.

Combining the positive aspects in water (TDS, hardness and pH as outlined) with an intelligent approach of filtering out harmful substances can result in drinking water that is beneficial and safe.

By carefully evaluating bottled water and home filtration systems, you will be able to select the type of water that fits your needs - both health-wise and dollar-wise.

10 • SUMMARY

We have come to the end of our analysis and recommendations on drinking water. One of the main points of emphasis has been to locate the beneficial, positive qualities in our drinking water. Which is the best water to drink? What should it contain?

To this end, we have located three significant factors - total dissolved solids, hardness and pH. By looking at the whole body of drinking water research in a comprehensive way, we have seen strong indications of what our 'ideal' drinking water should contain. The best water to drink should have:

- approximately 300 ppm (or mg/L) total dissolved solids (TDS)
- around 170 ppm (or mg/L) hardness (measured by CaCo3 - calcium carbonate)
- an alkaline pH - over 7.0

If we drank this type of water, we would have less heart disease and cancer mortality.

Numerous research studies presented indicate that in the United States, we could expect to have up to 15% less cardiovascular deaths (using hard water instead of soft water), 10 to 25% less cancer (if adequate TDS and silica are present) and an additional 10% reduction in cancer mortality if we stopped artificial fluoridation. Thus, if we drank the proper water, close to 300,000 annual deaths could be averted. Absolutely staggering!! And, if we removed chlorine before drinking our water, these figures would, no doubt, be substantially greater.

Changing the quality of your drinking water should be fairly easy with the information on home filtration units and bottled water. You will need to make sure your drinking water is close to the 'ideal' figures regarding total dissolved solids, hardness and pH. The choice is yours!

Drinking the correct type of water may be a missing and vital link in your overall health plan. Here's drinking to your health!

11 • UPDATE TO THE SECOND EDITION

Since the publication of Healthy Water For A Longer Life in July, 1984, several excellent research articles have been published on drinking water and its relationship to our health. The folowing comments briefly summarize these findings and as you will see greatly support the major recommendations given earlier as to what constitutes a healthy drinking water.

A report by the Oak Ridge National Laboratory found that the calcium and magnesium in hard water reduces the risks of heart attacks and strokes.(1)(2) This study compared the health records of 1,400 Wisconsion male farmers who drank water from wells on their own farms. The farmers who drank soft water suffered from heart disease. Whereas, the farmers who drank hard water were, for the most part, free of the problem. In many ways, these observations corroborate the findings of the British Regional Heart Study (see Chapter 1 for details).

Recent articles in the journal, Magnesium, have pointed out that studies on water magnesium levels alone as opposed to those on water hardness (Ca & Mg) have all shown a reverse correlation between cardiovascular mortality and magnesium level. Namely, the higher levels of magnesium in drinking water result in less heart disease problems.(3)(4) In the United States and South Africa, a 6 to 8 mg/L of waterborne magnesium would result in a 10% reduction in heart disease mortality. Magnesium is not only a vital nutrient for the heart and circulatory system but it is also an antagonist and inhibitor of lead and cadmium absorption. In addition, waterborne magnesium may have approximately one third better absorption than dietary magnesium.

A study in the region of Abruzzo, Italy, for the years 1969-1978 with a population of 600,000, found an inverse correlation between water hardness and mortality due to cardiovascular disease.(5) The CaCo3 levels in these regions ranged from 105.6 to 443.5 mg/L. Similar findings have been previously reported (see Chapter 1).

I want to conclude this update with what I consider one of the most major and significant pieces of research done on water and its impact on our health.

The ingestion of harmful chemicals from drinking water may not be the primary route to exposure.(6) Skin versus oral absorption rates for three volatile organic chemicals (toluene, ethylbenzene, styrene) using three different exposure amounts in adults and children were calculated. According to Dr. Halina Brown, the absorption rates for these toxic chemicals appears to be representative of chemical compounds commonly found in our drinking water.(7) The following summary chart is derived from this research.

TABLE 10

AVERAGE SKIN ABSORPTION VERSUS ORAL INGESTION

	Skin Absorption	Exposure Time	Oral Ingestion	Water Consumed Per Day
Adult bathing	63%	15 min.	27%	2 liters
Infant bathing	40%	15 min.	60%	1 liter
Child swimming	88%	1 hour	12%	1 liter
Overall average	64%		36%	

As you can see the rates of skin absorption are tremendous. Those of you with pools and hot tubs especially, take note! Keep in mind these calculations are based on skin absorption studies on the hand which is a much better barrier against harmful substances than other skin areas, which are more sensitive. This means the true absorption rates are significantly higher! And "ironically, highest doses resulted from water with the smallest concentration of the pollutant. In fact, diluted concentrations enter the skin far more easily than a pure chemical will . . . because most organics dehydrate skin making it less porous."(8)

The implications of this research is overwhelming! Obviously, having an acceptable point of use home filter or proper bottled drinking water is not adequate protection from harmful waterborne chemicals. What can we do?

Ideally, one should have a whole house filtration system installed to remove the organic volatile chemicals from bathing water. Yet, one of the problems is to properly evaluate the effectiveness of these units. In the past, different types of whole house filtration systems have been successfully used to remove a variety of minerals from the water, such as iron, etc. But, my recommendation for installing the system is for chemical removal, not mineral removal.

At this time, I am not aware of any independent comparison product test reports on whole house filtration systems. The consumer needs to be alert and apply the same principles used in evaluating drinking water filters (Chapter 9) to whole house filtration systems. Another concern I have is whether a household will need both a point of use drinking water filter and a whole house system or will a proper whole house filter by sufficient?

I realize these comments reveal the difficulties in evaluating whole house filtration systems but I believe the potential health benefits are well worth the effort. Skin absorption of harmful chemicals is a real issue. We need to face it and solve it!

Some concerned and observant writers have begun to recommend the installation of water filters on the shower head and bath tub faucets as a method of removing these harmful chemicals. However, I presently have my doubts with this approach. My main concern is whether these filters can effectively remove organic chemicals from <u>hot water.</u> With an increase in water temperature, the effectiveness of activated granulated carbon to remove volatile organic chemicals is greatly compromised. Interestingly, most manufacturers of drinking water filtation systems recommend only passing cold water through them. Whereas some units actually use hot water as a means to backflush or backwash the captured contaminants from the carbon. If you do purchase special units for the shower and bath, make sure the products can truly do what you want them to do - effectively remove organic contaminants from hot water.

Paying attention to the quality of our drinking water only is not enough. "You can now envision," Dr. Brown told me, "a situation where the total body burden of volatile chemicals will be distributed roughly one third from inhalation during showering, one third from oral ingestion and one third from washing/bathing. In effect, this easily doubles or triples our exposure to the harmful chemicals found in water."(9)

I believe we can still find and create a healthy water for drinking and bathing. We need to follow the guidelines set forth in this book regarding hardness, TDS and pH for drinking water, whether its bottled or filtered and in addition, install effective, whole house filtration systems for safe and clean bathing. By combining both approaches, we can begin to "drink" a healthy water for a longer life.

* FOOTNOTES

CHAPTER ONE: THE WATER STORY AND HEART DISEASE

(1) H.A. Schroeder, "Relation Between Mortality From Cardiovascular Disease and Treated Water Supplies," J. Am. Medical Assoc. (April 1960):98.

(2) G.W. Comstock, "Reviews and Commentary: Water Hardness and Cardiovascular Diseases," Am. J. Epidemiology 110(October 1979):377.

(3) L.C. Neri, et al., "Health Aspects of Hard and Soft Waters," J. Am. Water Works 67(August 1975):403-409.

(4) Idem, Comstock, p. 387.

(5) Ibid., p. 395-396.

(6) A.R. Sharrett, "The Role of Chemical Constituents of Drinking Water in Cardiovascular Diseases," Am. J. Epidemiology 110(October 1979):416.

(7) Ibid., p. 406.

(8) Ibid., p. 416.

(9) M.D. Crawford, M.J. Gardner and J.N. Morris, "Cardiovascular Disease and the Mineral Content of Drinking Water," British Medical Bull. 27(1971):23.

(10) H.I. Sauer, "Relationship of Water to Risk of Dying," in International Water Quality Symposium ed. by D.X. Manners, p. 76.

(11) Ibid., p. 77.

(12) A.R. Sharrett, et al., "Panel Discussion: The Relationship of Hard Water and Soft Water in CVD and Health," J. Environmental Pathology and Toxicology (1980): 125.

(13) V. Puddu, et al., "Drinking Water and Cardiovascular Disease," Am. Heart J. 99(April 1980):539-540.

(14) R. Masironi and A.G. Shaper, "Epidemiological Studies of Health Effects of Water from Different Sources," Annual Review of Nutrition 1(1981):396-397.

(15) A.G. Shaper, et al., "British Regional Heart Study: Cardiovascular Risk Factors in Middle-aged Men in 24 Towns," British Medical J. 283(July 1981):179-186.

(16) D.G. Greathouse and R.H. Osborne, "Preliminary Report on Nationwide Study of Drinking Water and Cardiovascular Diseases," J. Environmental Pathology and Toxicology 3(1980):65.

(17) National Research Council, Drinking Water and Health, (Washington, D.C.:National Academic Press, 1977), v. 1, p. 477.

CHAPTER TWO:
SODIUM, HYPERTENSION AND DRINKING WATER

(1) R.W. Tuthill and E.J. Calabrese, "Age as a Function in the Development of Sodium Related Hypertension," Environmental Health Perspectives 29(April 1979):35.

(2) H. Blackburn and R. Prineas, "Diet and Hypertension: Anthropology, Epidemiology, and Public Health Implications," Prog. Biochemical Pharmacology 19(1983):70.

(3) J. Arehart-Treichel, "Eating Your Way Out of High Blood Pressure," Science News 123(April 9, 1983):232-233.

(4) R.W.Tuthill and E.J. Calabrese, "Elevated Sodium Levels in the Public Drinking Water as a Contributor to Elevated Blood Pressure Levels in the Community," Archives of Environmental Health 34(July/August 1979):197.

(5) W.C. Willett, "Drinking Water Sodium and Blood Pressure: A Cautious View of the 'Second Look'," Am. J. of Public Health 71(July 1981):729-732.

(6) B.K. Armstrong, et al., "Water Sodium and Blood Pressure in Rural School Children," Archives of Environmental Health 37(July/August 1982):243.

(7) A. Hoffman, H.A. Valkenburg and G.J. Valkenburg, "Increased Blood Pressure in School Children Related to High Sodium Levels in Drinking Water," J. of Epidemiology and Community Health 34(1980):179.

(8) W.H. Hallenbeck, G.R. Brenniman and R.J. Anderson, "High Sodium in Drinking Water and Its Effect on Blood Pressure," Am. J. of Epidemiology 114(1981):817.

(9) Ibid.

(10) H.S. Faust, "Effects of Drinking Water and Total Sodium Intake on Blood Pressure," Am. J. of Clinical Nutrition 35(June 1982):1459.

(11) E.V.Ohanian and D.M. Cirolia, "Sodium in Drinking Water as an Etiological Factor in Hypertension," Water Technology (October 1983):32.

(12) Ibid.

(13) J.S. Robertson, J.A. Slattery and V. Parker, "Water Sodium, Hypertension and Mortality," Community Medicine 1(1979):295.

(14) H.A. Schroeder, "Relation Between Mortality From Cardiovascular Disease and Treated Water Supplies, J. Am. Medical Assoc. (April 1960):103-104.

(15) H.I. Sauer, "Relationship of Water to the Risk of Dying," in International Water Quality Symposium ed. by D.X. Manners, p. 78.

(16) D.G. Greathouse and R.H. Osborne, "Preliminary Report on Nationwide Study of Drinking Water and Cardiovascular Diseases," J. of Environmental Pathology and Toxicology 3(1980):65.

(17) E.J. Calabrese, et al., "The Role of Elevated Levels of Sodium in Diet and Drinking Water on the Development of Hypertension in Animal Models and Humans," J. of Environmental Pathology and Toxicology 3(1980):143-144.

(18) R. Masironi and A.G. Shaper, "Epidemiological Studies of Health Effects of Water From Different Sources." Annual Review of Nutrition 1(1981):382-383.

(19) Idem, Calcabrese, p. 143.

(20) Idem, Greathouse, p. 72.

(21) Idem, Masironi, p. 382-383.

(22) Idem, Blackburn, p. 42.

CHAPTER THREE: THE WATER STORY AND CANCER

(1) S.S. Epstein and M. Zavon, "Is There a Threshold for Cancer," in International Water Quality Symposium ed. by D.X. Manners, p. 61.

(2) Ibid., p. 62.

(3) Ibid., p. 59.

(4) T. Page, et al., "Drinking Water and Cancer Mortality in Louisiana," Science (July 1976):55.

(5) A.C. Burton and F. Cornhill, "Correlation of Cancer Death Rates with Altitude and with the Quality of Water Supply of 100 Largest Cities in the United States," J. Toxicology and Environmental Health 3(1977):465-478.

(6) Ibid., pp. 476-477.

(7) A.C. Burton, J.F. Cornhill and B. Canham, "Protection From Cancer by 'Silica' in the Water Supply of U.S. Cities," J. Environmental Pathology and Toxicology 4(1980):32.

(8) Idem, Burton, "Correlation of Cancer Deaths . . ." p. 470.

(9) Idem, Burton, "Protection From Cancer . . . " p. 32.

(10) H.A. Schroeder, "Relation Between Mortality From Cardiovascular Disease and Treated Water Supplies," J. Am. Medical Assoc. (April 1960):103.

(11) G. Lolli, L.A. Greenberg and D. Lester, "The Influence of Carbonated Water on Gastric Emptying," New England J. Medicine 246(1951):490-492.

(12) L.A. Greenberg and J.M. Turner, "Effect of Carbonated Water on Gastric Acid Secretion," New England J. Medicine 253(1955):105-106.

(13) B.J.A. Haring and B.C.J. Zoeteman, "Corrosiveness of Drinking Water and Cardiovascular Disease Mortality," Bull. Environmental Contaminant Toxicology

(14) Idem, Burton, "Protection From Cancer . . ." p. 31.

(15) Ibid., p. 39.

(16) Ibid., p. 40.

(17) Ibid.

(18) H.I. Sauer, "Relationship of Water to the Risk of Dying," in International Water Quality Symposium ed. by D.X. Manners, p. 77.

(19) B. Jansson, "Seneca County, New York: An Area With Low Cancer Morality Rates," Cancer 48(1981):2542.

(20) Ibid., p. 2546.

(21) J. Thouez, Y. Beauchamp and A. Simard, "Cancer and the Physicochemical Quality of Drinking Water in Quebec," Social Science and Medicine 15D(1981):218.

CHAPTER FOUR: FLUORIDATION AND CANCER

(1) D. Burk, "Fluoridation: A Burning Controversey," Bestways (April 1982):44.

(2) Ibid., p. 42.

(3) C. Bukro, "Judge's Ruling Muddles Illinois Drinking Water," Health Freedom News (April 1983):28-29.

(4) J.A. Yiamouyiannis, Everything You Wanted to Know About Fluoridation - But Were Afraid to Ask: A Discovery Deposition, (Monrovia, CA: National Health Federation, 1977), p. 54.

(5) A.F. Furman, "Fluoridation Boon or Bane," Health Express (October 1982):55.

(6) G.L. Waldbott, et al., Fluroidation: The Great Dilemma, (Lawrence, KS:Coronado Press, 1978), p. 219.

(7) Ibid., p. 393.

(8) Ibid., P. 392.

(9) J. Yiamouyiannis, Lifesavers Guide to Fluoridation: Risk/Benefits Evaluated in this 1982 Question and Answer Report, (Delaware, Ohio: Safe Water Foundation, 1982), p. 4.

(10) Idem, Burk, p. 44.

(11) W.A. Price, "Race Decline and Race Regeneration," J. Am. Dental Assoc. 28(1941): 550.

(12) Idem, Waldbott, p. 192.

(13) Idem, Yiamouyiannis, Everything You Wanted . . . p. 20.

(14) Ibid., p. 6.

(15) Ibid., p. 25

(16) Ibid., p. 28.

(17) D. Burk, Personal Communication, May 1983.

(18) Idem, Burk "Fluoridation: A Burning Controversey," p. 42.

(19) Ibid., p. 44.

CHAPTER FIVE: ASBESTOS AND CANCER

(1) I.J.Selikoff, "Asbestos in Water," in *International Water Quality Symposium* ed. by D.X. Manners, pp. 66-70.

(2) Ibid., p. 66.

(3) Ibid., pp. 66-70.

(4) S.S. Epstein and M. Zavon, "Is There a Threshold for Cancer," in *International Water Quality Symposium* ed. by D.X. Manners, p. 61.

(5) E.E. Sigurdson, et al., "Cancer Morbidity Investigations: Lessons From the Duluth Study of Possible Effects of Asbestos in Drinking Water," *Environmental Research* 25(1981):50-61.

(6) K.W. Donsbach and M. Walker, *Drinking Water*, (Huntington Beach, CA:International Institute of Natural Health Sciences, 1981), p. 9.

(7) P.M. Conforti, et al., "Asbestos in Drinking Water and Cancer in the San Francisco Bay Area: 1969-1974 Incidence," *J. Chronic Diseases* 34(1981):211.

CHAPTER SIX: CHLORINATION: A LINK BETWEEN HEART DISEASE AND CANCER

(1) R. Passwater, *Supernutrition for Healthy Hearts*, (New York: Jove, 1978), p. 312.

(2) J.M. Price, *Coronaries/Cholesterol/Chlorine*, (New York: Pyramid, 1969), p. 49.

(3) Ibid., pp. 84-85.

(4) Ibid., p. 16.

(5) Ibid., p. 18.

(6) Ibid., pp. 18-19.

(7) Ibid., p. 22.

(8) Ibid., p . 23.

(9) Ibid., p. 53.

(10) "Cancer and Chlorinated Water," *Lancet* (May 23, 1981):1142.

(11) T.H. Maugh II, "New Study Link Chlorination and Cancer," *Science* 211(February 13, 1983):694.

(12) J.R. Wilkins, et al., "Organic Chemical Contaminants in Drinking Water and Cancer," Am. J. Epidemology 110(October 1979):438.

(13) Ibid., p. 423.

(14) Ibid.

(15) Ibid.

(16) Ibid., p. 424.

(17) "Water - A Clear and Present Danger," ABC News (August 5, 1983:Show #119 Transcript):1.

(18) K.W. Donsbach and M. Walker, Drinking Water, (Huntington Beach, CA:International Institute of Natural Health Sciences, 1981), p. 9.

(19) T. Page, et al., "Drinking Water and Cancer Mortality in Louisiana," Science 193 (July 1976):55.

(20) M.S. Gottlieb et al., "Cancer and Drinking Water in Louisiana: Colon and Rectum," International J. Epidemology 10(June 1981):117-125.

(21) G.L. Carlo and C.J. Mettlin, "Cancer Incidence and Trihalomethane Concentrations in a Public Water System," Am. J. Public Health 70(May 1980):524.

(22) J.R. Wilkins and G.W. Comstock, "Source of Drinking Water at Home and Site-Specific Cancer Incidence in Washington County, Maryland," Am. J. Epidemiology 114(1981):178.

(23) Ibid., pp. 188-189.

(24) Idem, Donsbach, pp. 11-12.

(25) A.C. Anderson, et al., "A Brief Review of the Current Status of Alternatives to Chlorine Disinfection of Water," Am. J. Public Health 72(1982):1290.

(26) Ibid., p. 1291.

(27) Idem, Passwater, pp. 312-313.

CHAPTER SEVEN: THE WATER STORY AND ANIMAL EXPERIMENTS

(1) P.T. McCauley and P.T. Bull, "Experimental Approaches to Evaluating the Role of Environmental Factors in the Development of Cardiovascular Disease," J. Environmental Pathology and Toxicology 4(September 1980):27.

(2) R.F. Bolgman and S.F. Lightsey, "Effects of Synthesized Hard Water and of Cadmium in the Drinking Water Upon Lipid Metabolism and Cholelithiasis in Rabbits," Am. J. of Veterinary Research 43(August 1982):1432.

(3) H.M. Perry, et al., "Possible Influence of Heavy Metals in Cardiovascular Disease: Introduction and Overview," J. Environmental Pathology and Toxicology 3(1980) :195.

(4) Ibid., p. 196.

(5) C. Elinder, et al., "Water Hardness in Relation to Cadmium Accumulation and Microscopic Signs of Cardiovascular Disease in Horses," Archives of Environmental Health 35(March/April 1980):83.

(6) Ibid., p. 84.

(7) Ibid., p. 81 and 83.

(8) Idem, Borgman, pp. 1432-1435.

(9) J.M. Price, Coronaries/Cholesterol/Chlorine, (New York:Pyramid, 1969), pp. 64-70.

(10) J.B. Neal and M. Neal, "Effect of Hard Water and MgSO4 on Rabbit Atherosclerosis," Archives of Pathology 73(May 1963):58.

(11) Ibid., p. 60.

(12) Ibid., p. 58.

(13) R.S. Ingols and T.F. Craft, "Analytical Notes - Hard- vs. Soft-Water Effects on the Transfer of Metallic Ions from Intestine," J. Am. Water Works Assoc. 68(April 1976):210.

(14) "Calcium, Chlorine and Heart Disease Linked," Science News (August 13, 1983):103.

(15) Ibid.

(16) Ibid.

(17) Ibid.

(18) Ibid.

CHAPTER EIGHT:
DE-MINERALIZED WATER:
IS IT HEALTHY TO DRINK?

(1) B.L. Morales and J. Clark, "Longevity, Plus-I, II, III," Organic Consumer Report 62(September 14, 21 and 28, 1982):No. 37, 38 and 39.

(2) Ibid.

(3) W. Mertz, "Chromium, Selenium and Other Essential Trace Elements," in International Water Quality Symposium ed. by D.X. Manners, p. 41.

(4) M.J. Dauncey and E.M. Widdowson, "Urinary Excretion of Calcium, Magnesium, Sodium, and Potassium in Hard and Soft Water Areas," Lancet(April 1, 1972): 712.

(5) Ibid., p. 714.

(6) M.D. Crawford, M.J. Gardner and J.N. Morris, "Cardiovascular Disease and the Mineral Content of Drinking Water," British Medical Bull. 27(1971):21-24.

(7) A.R. Sharrett, et al., "Panel Discussion: The Relationship of Hard Water and Soft Water in CVD and Health," J. Environmental Pathology and Toxicology 4 (1980):113-141.

(8) A.C. Burton, et al., "Protection From Cancer by 'Silica' in the Water-Supply of U.S. Cities," J. Environmental Pathology and Toxicology 4(1980):31-40.

(9) A.W. Voors, "Minerals in the Municipal Water and Atherosclerotic Heart Death," Am. J. Epidemiology 93(1971):259-266.

(10) L.C. Neri, et al., "Is There a Water Factor: A Case for Magnesium," in International Water Quality Symposium ed. by D.X. Manners, pp. 73-75.

(11) L.C. Neri, et al., "Health Aspects of Hard and Soft Waters," J. Am. Water Works Assoc. 67(August 1975):403-409.

(12) H.A. Schroeder, "The Role of Trace Elements in Cardiovascular Disease," Medical Clinics of North America 58(March 1974):381-396.

(13) E.B. Dawson, et al., "Shall We Add Lithium to Drinking Water?" in International Water Quality Symposium ed. by D.X. Manners, pp. 47-52.

(14) W. Mertz, "Chromium, Selenium and Other Essential Trace Elements," in International Water Quality Symposium ed. by D.X. Manners, pp. 41-43.

(15) E.B. Dawson, et al., "Relationship of Metal Metabolism to Vascular Disease Mortality Rates in Texas," Am. J. Clinical Nutrition 31(July 1978):1188-1197.

(16) J. Sorenson, "Personal Communication," November 3, 1983. Phone conversation.

(17) Ibid.

(18) D. Ashmead, Chelated Mineral Nutrition, (Huntington Beach, CA:International Institute of Natural Health Sciences, 1981), pp. 17-18.

(19) Ibid., p. 18.

(20) D. Churchill, et al., "Drinking Water Hardness and Urolithiasis," Annals of Internal Medicine 88(April 1978):513-514.

(21) R. Sierakowski, B. Finlayson and R. Landes, "Stone Incidence as Related to Water Hardness in Different Geographical Regions of the United States," Urological Research 7(1979):157-160.

(22) G. Bryant and R. Jordan, "Effect of Different Cooking Waters on Calcium Content of Certain Vegetables," Food Research (1948):308-314.

(23) B.S.A. Haring and W. Van Delft, "Changes in the Mineral Composition of Food as a Result of Cooking in 'Hard' and 'Soft' Waters," Archives of Environmental Health 36(January/February 1981):33-35.

(24) "Your Government and Your Health," Prevention (September 1983):169.

(25) L. Clark, The Best of Linda Clark: Some Unusual Approaches to Health, (New Canaan, CT:Keats, 1976), pp. 32-46.

CHAPTER NINE: WHICH WATER IS BEST TO DRINK? AN EVALUATION OF BOTTLED WATER AND HOME FILTRATION SYSTEMS

(1) "Water's Rise: A Sales Torrent," Business Week (January 11, 1982):99.

(2) J. Studlick and R. Bain, "Bottled Water: Expensive Ground Water," Water Well J. (July 1980):75-79.

(3) Ibid., p. 75.

(4) U.S. Code of Regulations, Title 21: Part 103: Subpart B: Standards of Quality, (Washington GPO, 1983), Paragraph 103.35, 52-54.

(5) EPA, Bottled Water Study: A Pilot Survey of Water Bottlers and Bottled Water, (Washington GPO, 1972), 1-2.

(6) U.S. Code of Regulations, Title 21: Part 129: Processing and Bottling of Bottled Drinking Water, (Washington GPO, 1983), 136.

(7) R. Nacewicz, Personal Communication, FDA, (Washington, D.C.), November 1, 1983. Phone Conversation.

(8) W.F. Deal, Personal Communication, International Bottled Water Association, (Alexandria, VA), November 1, 1983. Phone Conversation.

(9) EPA, "Bottled Water," (A one page statement on bottled water. Not dated), Criteria and Standards Division. Office of Drinking Water.

(10) D.J. Edelman, Inc. "Information Packet," New York, 1982.

(11) "Suspected Carcinogen Found in Water," Health Freedom News, (September 1982):16.

(12) Idem, Business Week, p. 99.

(13) R.T. Williams, "Implied Claims Verification Testing-Water Filters/Purifiers - 800530," General Ecology, Inc. 1980.

(14) EPA, "Home Drinking Water Treatment Units Contract," Criteria and Standards Division, Office of Drinking Water. Phase 1 & 2. July, 1980, Phase 3, March 1982.

(15) L.K.L. Au, (Water Supply Section, EPA Office, San Francisco, CA), Personal Communication, January 7, 1983. Letter.

(16) "Water Filters," Consumer Reports, (February 1983):72-73.

(17) Idem, EPA, "Home Drinking Water . . . " Phase 1:3.

(18) Ibid.

(19) EPA, "Home Drinking Water Treatment Units Contract," Criteria and Standards Division, Office of Drinking Water. Phase 3. March, 1982:3.

(20) W.H. Beauman, "Purifiers," Everpure Bulletin, (September 13, 1983):1.

(21) Ibid.

(22) Idem, "Water Filters," Consumer Reports:102.

(23) B. Chambers, Personal Communication, Chief Chemist, Environmental Engineering Lab, (San Diego, CA), November 3, 1983. Interview.

(24) Ibid.

(25) EPA, "National Interim Primary Drinking Water Regulations," Federal Register, (December 24, 1975; July 9, 1976; November 29, 1979; March 11, 1980; August 27, 1980; March 3, 1982; March 12, 1982):Sections 141.1-141.42.

(26) EPA, "Drinking Water and Health: Recommendations of the National Academy of Sciences," Federal Register 42(July 11, 1977):35766.

CHAPTER ELEVEN: UPDATE TO THE SECOND EDITION

(1) E.A. Zeighami, et al, "Drinking Water Inorganics and Cardiovascular Disease: A Case-Control Study Among Wisconsin Farmers," pp. 135-158 in Inorganics in Drinking Water and Cardiovascular Disease edited by E.J. Calabrese, R.W. Tuthill, L. Condie, 1985. Princeton, NJ.

(2) "Hard Water May Cut Risk of Having a Heart Attack," U.S. Water News, (November 1985): p. 14.

(3) J. Durlach, M. Bara, A. Givet-Bara, "Magnesium Level in Drinking Water and Cardiovascular Risk Factor: A Hypothesis," Magnesium 4(1985):5-15.

(4) J.R. Marier, L.C. Neri, "Quantifying the Role of Magnesium in the Interrelationship Between Human Mortality/Morbidity and Water Hardness," Magnesium 4(1985):53-59.

(5) V. Leoni, L. Fabiani, L. Ticchiarelli, "Water Hardness and Cardiovascular Mortality Rate in Abruzzo, Italy," Archives of Environmental Health 40(1985):274-278.

(6) H.S. Brown, D.R. Bishop, C.A. Rowan, "The Role of Skin Absorption as a Route of Exposure for Volatile Organic Compounds (VOCs) in Drinking Water," American Journal of Public Health 74(May 1984):479-484.

(7) H.S. Brown, "Personal Communication," July 16,1986. Phone Conversation.

(8) "Water Pollutants - Exposure by Skin," Science News, (May 19, 1984):309.

(9) H.S. Brown, "Personal Communication," July 16, 1986. Phone Conversation.

APPENDIX ONE

BOTTLED WATER ANALYSIS

Adapted from, "Bottled Waters - Expensive Ground Water," Studlick and Bain, Ground Water, 18(July/August 1980):340-345. (All concentrations in mg/L unless otherwise noted.)

	Specific Conductance (µmhos)	Dissolved Solids	pH	Hardness Total (as CaCo3)
Absorpure	470	180	8.0	180
Apollinaris	2800	2200	6.5	470
Borden Polar	330	180	7.3	110
Calistoga	700	NA	6.0	56
Crodo	1800	NA	5.7	NA
Deer Park	70	60	6.8	20
Distillata	600	340	7.8	170
Magnetic Springs	55	80	6.2	24
Mountain Valley	300	120	7.8	190
Perrier	650	560	6.1	330
Poland	205	160	7.6	76
Ramlosa	800	600	6.3	8.0
Saratoga Vichy	3000	2100	6.3	280
Vichy Saint-Yorre	5400	4300	6.9	64

NOTE:

The water analyses presented in Appendix 1, 2 and 3 do not specify which waters are processed and which are natural. Also, the reader should know that there are inconsistencies between the charts, e.g. the analysis for total dissolved solids may vary in each article for the same product. However, taken as a group, this data should greatly assist in the seletion of a bottled water that fulfills the criteria outlined in Chapter Nine, Which Water Is Best To Drink?

APPENDIX TWO

BOTTLED WATER AND TOTAL DISSOLVED SOLIDS

Adapted from, "The Selling of H20," Consumer Reports, September, 1980:531-538.

I. STILL WATERS	Total Dissolved Solids (ppm)
Arrowhead Mountain Spring	99
Bonniebrook Spring	30
Borden Polar Spring	57
Carolina Mountain	55
Deep Rock Artesian Fresh Drinking	600
Deer Park 100% Spring	29
Evian Natural Spring	330
Fiuggi Natural Spring	120
Great Bear Natural Spring	13
Hinckley and Schmitt Natural Spring	440
Mountain Spring	52
Mountain Valley	195
Poland Spring Pure Natural Mineral	125
Sparkletts Crystal-Fresh Drinking	140
II. SPARKLING WATERS	
Apollinaris Natural Mineral	2250
Bartlett Mineral Spring Sparkling	1950
Black Forest Naturally Sparkling Geniune Mineral	905
Calistoga Sparkling Mineral	540
Calso Mineral	5100
Canada Dry Club Soda	536
Canada Dry Seltzer Pure Sparkling	135
Deer Park Sparkling 100% Spring	24
Ferrarelle Naturally Sparkling Mineral	1400
Gerolsteiner Sprudel Natural Mineral	1750
Le-Nature's Crystal Clear Mineral	365
Montclair Sparkling Natural Mineral	200
Perrier Naturally Sparkling Mineral	545
Peters Val Naturally Sparkling Mineral	1350
Poland Spring Sparkling Pure Natural Mineral	99
Safeway Bel-Air Sparkling Mineral	565
San Pellegrino Natural Sparkling Mineral	1100
á Sante Napa Valley Mineral Sparkling	645
Saratoga Naturally Sparkling Mineral	445
Schweppes Sparkling Mineral	365
Sheffield's 02 Sparkling Spring	73
Vichy Celestins Naturally Alkaline Mineral	3400
Vitteloise Natural Spring	465

APPENDIX THREE

MINERAL, SPRING AND VICHY WATER ANALYSIS

Adapted from, "1982 Report on Bottled Water and Bottled Water Substitutes," by the Suffolk County Department of Health Services, New York.

	Specific Conductivity (µmhos)	pH	Hardness CaCo3 (mg/L)
I. MINERAL WATER			
Apollinaris	>3000	5.8	636
Badoit	1800	6.0	560
Coralba Natural	370	8.2	NA
Evian	550	7.4	NA
Ferrarelle	>2200	5.9	NA
Fiuggi	143	6.3	NA
Montclair	310	4.9	NA
Perrier	820	5.2	NA
Peters Vol	>1750	5.5	NA
Poland Spring	76	4.1	NA
Poland Spring - Non-Efferv.	42	6.5	NA
Rambosa	850	5.5	NA
San Pellegrino	1400	5.2	NA
San Rita	142	6.1	NA
Saratoga Geyser	>6600	6.1	NA
Saratoga Natural	725	5.3	NA
Staatl Fachingen	>3000	6.3	NA
White Rock	177	4.1	NA
II. SPRING WATER			
Adirondock	56	7.4	NA
Ann Page	78	6.5	NA
Blue Rock	27	6.2	NA
Bonniebrook	290	6.6	NA
Contrexeville	>2100	7.3	1560
Covered Bridge	60	6.9	NA
Deer Park Sparkling	325	4.0	NA
Deer Park 100%	46	6.5	NA
Edwards/Finast	240	7.5	NA
Foodtown	63	6.4	NA
Forest Springs	550	7.7	NA
Grand Union	46	6.7	NA
Great Bear (Sassoonan)	33	6.7	NA
Great Bear (Shamrock)	280	7.7	NA
Holly Spring	320	8.0	NA
Jaggers	135	4.4	NA
Light Rock	400	8.3	NA
Mount Laurel Natural	95	7.2	NA
Mount Laurel Sparkling	240	4.8	NA
Mountain Spring	48	6.2	NA
Mountain Stream	129	7.2	NA
Mountain Valley	350	7.7	NA
No Frills	70	6.2	NA

APPENDIX THREE (Continued)

	Specific Conductivity (mhos)	pH	Hardness CaCo3 (mg/L)
Old Fashioned	275	4.8	NA
Pathmark Natural	235	6.2	NA
Polar Spring	53	5.9	NA
Rite-Aid	65	7.0	NA
Saratoga Natural	760	5.2	NA
Shamrock Sparkling	153	NA	NA
Shoprite Natural	53	6.2	NA
Spring House	32	6.2	NA
Vitteloise	650	5.1	NA
Waldbaums	68	6.8	NA
White Rose	54	6.3	NA
III. VICHY WATER			
British American	>3900	5.2	NA
Celestins	>5200	6.4	NA
Saint Yoree	>6800	6.4	NA

APPENDIX FOUR

Trihalomethane (THM) and Non-Purgeable Total Organic Carbon (NPTOC) Reductions, EPA/GSRI Contract

"Trihalomethan Reductions" from: EPA, Home/Water Units Contract, Second Phase, July, 1980.

INFORMATION ON UNITS TESTED				AVERAGE REDUCTIONS	
NAME	Test Life (gal)	Est. Ave Contact (sec)	Weight of Carbon (gms)	THM (%)	NPTOC (%)
PHASE II - Commercial					
FAUCET-BYPASS					
1. Aquaguard, Model AGT-31 Cartridge T-3XL	500	3.4	51	43	12
2. Concept Bacteriostatic Home Water Filter	40	1.6	NA	16	18
3. Filter Fresh Model FF-1	1200	4.8	59	6	6
4. Hurley Town and Country	4000	36.5	895	69	31
5. Water Washer, Countertop Model 1000	1000	10.9	70	41	11
POUR-THROUGH					
6. Filbrook Pour-Thru Activated Carbon	1000	43.8	97	40	14
7. Puriton Bacteriostatic Drinking Water Treatment Unit	1000	14.1	30	21	6
STATIONARY					
8. AMF-Cuno Housing 1M Cartridge AP-117	3000	3.9	395	34	7
9. Filterite, Model 1 PC Cartridge 1C-9	3000	1.2	NA	18	8
10. Fulfo Water Filter Model WC-10	3000	2.6	208	15	11
11. Keystone Model 3121 Housing with Model 310 Cartridge	3000	2.3	NA	21	9

APPENDIX FOUR (Continued)

INFORMATION ON UNITS TESTED

NAME	Test Life (gal)	Est.Ave Contact (sec)	Weight of Carbon (gms)	AVERAGE REDUCTIONS THM (%)	NPTOC (%)
LINE-BYPASS					
12. Aquacell Bacteriostatic Water Treatment Unit	2000	12.4	417	86	23
13. Aqualux Water Processor Model CB-2	2000	35.2	1150	98	23
14. Argenion Bacteriostatic Water Treatment Unit, Model I	2000	8.1	146	23	NA
15. Continental Water Filter Model 350	720	185	3402	99	87
16. Culligan Super Guard Model SG-2	4000	39	1708	89	28
17. Everpure, Model QC4-THM	1000	43	765	99	55
18. Mariner Renaturalizer Water Units	3000	31.4	560	47	21
19. Polarisdynamic Water Unit	2500	26.6	527	61	18
20. Purogen Water Detoxifier	2500	11.6	859	38	6
21. Seagull IV	1600	14.8	300	70	30
22. Ultrapure Bacteriostatic	3000	28.0	1103	40	20
23. Waterco, Model AS-5	3000	5.0	495	25	NA
OTHER					
24. Wunderbar Portable Water Cleaner-Filter	200	0.9	NA	4	NA

APPENDIX FOUR (Continued)

INFORMATION ON UNITS TESTED

NAME	Test Life (gal)	Est.Ave Contact (sec)	Weight of Carbon (gms)	AVERAGE REDUCTIONS THM (%)	AVERAGE REDUCTIONS NPTOC (%)
PHASE I - Commercial					
FAUCET-BYPASS					
25. Instapure Model FI-C	200	1.6	27	24	11
FAUCET-NO BYPASS					
26. Mini Aqua Filter	200	0.9	16	6	2
POUR-THROUGH					
27. H20K Portable Drinking Water Treatment Unit	2000	15	94	19	10
STATIONARY					
28. Sears Taste and Odor Filter	3420	4.5	398	46	12
LINE-BYPASS					
29. Aqualux Water Processor Model HB	2000	23	575	45	28
30. System I Water Processor Model SY 1-34	2500	46	1120	43	20
PHASE II - Experimental-Unique Resin					
31. Rohm and Haas Ambersorb XE-340	3500	50.6	2353	93	6

APPENDIX FIVE

TABLE ONE

Summary Performance: Halogenated Organics Removal During Rated Lifetime for Line Bypass, Faucet-Mount, Stationary, and Pour-Through Units in the Ground Water Study
"Halogenated Organics Reduction," from EPA, Home Water Treatment Units Contract, Third Phase, March 1982.

Unit	Rated Capacity (gallons/liters)	Average Influent Concentration (ppb)	Total Influent Load (g)	Amount Removed By Activated Carbon (g)	(%)
LINE BYPASS					
Aqualux CB-2	2000/7571	132	1.00	0.99	99
Continental Model 350	720/2725	134	0.37	0.37	99
Culligan Model SG-2	4000/15142	144	2.18	2.16	99
Everpure QC4-THM	1000/3785	158	0.60	0.59	99
Seagull IV	1000/3785	158	0.60	0.58	97
Aquacell	2000/7571	132	1.00	0.97	97
FAUCET-MOUNT					
Hurley Town and Country	4000/15142	113	2.17	2.10	97
Water Washer Model 1000	1000/3785	128	0.48	0.36	76
POUR-THROUGH					
Filbrook*	300/1136	143	0.27	0.25	94
STATIONARY					
Sears Taste and Odor	3000/11356	152	1.73	1.60	90

*Filbrook was tested for 500 gallons but results are adjusted to 300 gallons as being more representative of claimed lifetime.

APPENDIX FIVE (Continued)

TABLE TWO

Range of Percent Specific Halogenated Organic (HO) Reduction for Line Bypass Faucet-Mount, Stationary, and Pour-Through Units in the Ground Water Study

UNIT	Rated Capacity (gallons/liters)	Range of Average Percent HO Reduction (Beginning - Ending)			
		1	2	3	4
LINE BYPASS					
Aqualux CB-2	2000/7571	99-99	98-96	99-99	99-99
Continental Model 350	720/2725	99-99	99-99	99-99	99-99
Culligan Model SG-2	4000/15142	99-98	99-98	99-99	99-99
Everpure QC4-THM	1000/3785	99-99	95-99	99-99	99-99
Seagull IV	1000/3785	98-95	98-97	98-97	98-97
Aquacell	2000/7571	99-93	99-95	99-98	99-97
FAUCET-MOUNT					
Hurley Town and Country	4000/15142	99-93	97-94	99-98	99-99
Water Washer Model 1000	1000/3785	96-40	95-55	95-70	99-62
POUR-THROUGH					
Filbrook*	300/1136	99-72	98-82	95-94	99-98
STATIONARY					
Sears Taste and Odor	3000/11356	98-70	98-80	99-96	99-98

1 1,1,1-trichloroethane
2 carbon tetrachloride
3 trichloroethylene
4 tetrachloroethylene
* Filbrook was tested for 500 gallons but results are adjusted to 300 gallons as being more representative of claimed lifetime.

APPENDIX FIVE (Continued)

TABLE THREE

Range of Percent Specific Halogenated Organic (HO) Reduction for Line Bypass Faucet-Mount, Stationary, and Pour-Through Units in the Surface Water Study

UNIT	Rated Capacity (gallons/liters)	Range of Average Percent HO Reduction (Beginning - Ending)		
		1	2	3
LINE BYPASS				
Aqualux CB-2	2000/7571	99-90	99-54	99-89
Continental Model 350	720/2725	95-95	85-95	99-99
Culligan Model SG-2	4000/15142	77-89	99-45	95-83
Everpure QC4-THM*	1000/3785	99-99	99-99	99-99
Seagull IV*	1000/3785	99-99	99-99	99-98
Aquacell	2000/7571	99-96	99-80	99-89
FAUCET-MOUNT				
Hurley Town and Country	4000/15142	99-92	99-50	99-79
Water Washer Model 1000	1000/3785	99-60	50-20	96-60
POUR-THROUGH				
Filbrook**	300/1136	99-75	99-40	99-45
STATIONARY				
Sears Taste and Odor	3000/11356	79-88	51-30	78-35

1 p-Dichlorobenzene (C6H4Cl2)
2 Hexachlorobenzene (C6Cl6)
3 Chlordane (Technical grade, 60%)

* Units plugged prematurely
** Filbrook was tested for 500 gallons but results are adjusted to 300 gallons as being more representative of claimed lifetime.

APPENDIX SIX

NATIONAL INTERIM PRIMARY DRINKING WATER REGULATIONS

MAXIMUM CONTAMINANT LEVELS

I. COLIFORM BACTERIA

Coliform Method	Per Month	Less Than 20 Samples Per Month	More Than 20 Samples Per Month
A. MFT* 100 ml Standard Portions	Not to exceed 1/100 ml as the Arithmetic Mean	4/100 ml in more than one sample	4/100 ml in more than 5% of samples
B. FTM** 10 ml Standard Portions	Not to be present in more than 10% of portions	3 or more portions in more than 1 sample	3 or more portions in more than 5% of samples
		Less Than 5 Samples Per Month	More Than 5 Samples Per Month
C. FTM 100 ml Standard	Not to be present in more than 60% of portions	5 portions in more than 1 sample	5 portions in more than 20% of samples

* MFT - membrane filter technique
** FTM - fermentation tube method

The number of samples to be collected per month is dependent upon the service population.

The above MCLs are applicable to community and non-community water systems.

II. TURBIDITY

The MCL for turbidity in drinking water is one (1) turbidity unit (TU) as determined by one/day monthly average for surface supplies. Five (5) or fewer TU may be allowed if the higher turbidity does not do any of the following:

A. Interfere with disinfection.
B. Prevent maintenance of an effective disinfectant agent throughout the distribution.
C. Interfere with microbiological determinations.

The above MCL is applicable to both community and non-community water systems using surface water sources in whole or part.

III. INORGANICS

CONTAMINANT	LEVEL (mg/L)
Arsenic	0.05
Barium	1.00
Cadmium	0.010
Chromium	0.05
Fluoride	*
Lead	0.05
Mercury	0.002
Nitrate (as N)	10.00
Selenium	0.01
Silver	0.05

* The MCL for fluoride is related to the annual average of the maximum daily air temperature.

TEMPERATURE (F°)	LEVEL (mg/L)
53.7 and below	2.4
53.8 - 58.3	2.2
58.4 - 63.8	2.0
63.9 - 70.6	1.8
70.7 - 79.2	1.6
79.3 - 90.5	1.4

All the above MCLs, except nitrate, apply only to community water systems. The nitrate level applies to both community and non-community water systems.

IV. ORGANICS

CONTAMINANT	LEVEL (mg/L)
Endrin	0.0002
Lindane	0.004
Methoxychlor	0.1
Toxaphene	0.005
2,4-D	0.1
2,4,5-TP Silvex	0.01
TTHM	0.10

All the above apply only to community surface water systems.

V. RADIONUCLIDES

CONTAMINANT	PICOCURIES/l
Gross Alpha Activity	15
Radium-226 + Radium-228	5

All the above apply only to community water systems.

APPENDIX SEVEN

FDA BOTTLED WATER STANDARDS
TITLE 21 - U.S. CODE OF REGULATIONS 1983

Bottled water does not include mineral water. Any bottled water that does not exceed these standards must be so labeled.

(A) MICROBIOLOGICAL QUALITY

(1) Multiple-tube fermentation method: Not more than one of the analytical units in the sample shall have a most probable number (MPN) of 2.2 or more coliform organisms per 100 milliliters and no analytical unit shall have an MPN of 9.2 or more coliform organisms per 100 milliliters; or:

(2) Membrane filter method: Not more than one of the analytical units in the sample shall have 4.0 or more coliform organisms per 100 milliliters and the arithmetic mean of the coliform density of the sample shall not exceed one coliform organism per 100 milliliters.

(B) PHYSICAL QUALITY

(1) The turbidity shall not exceed 5 units.

(2) The color shall not exceed 15 units.

(3) The odor shall not exceed threshold order No. 3.

(C) CHEMICAL QUALITY

Substance - Concentration in Milligrams per Liter

Arsenic	0.05
Barium	1.0
Cadmium	0.01
Chloride	250.0
Chromium	0.05
Copper	1.0
Iron	0.3
Lead	0.05
Manganese	0.05
Mercury	0.002
Nitrate (N)	10.0
Phenols	0.001
Selenium	0.01
Silver	0.05
Sulfate	250.0
Total Dissolved Solids	500.0
Zinc	5.0

APPENDIX SEVEN (Continued)

(C) CHEMICAL QUALITY (Continued)

Organics

Endrin (1,2,3,4,10,10-hexachloro-6,7,-epoxy-1,4,4a,5,6,7,8,8a-Octa-hydro-1,4-endo, endo-5,8-aimethano naphthalene)	0.0002
Lindane (1,2,3,4,5,6,-nexachloro-cyclohexane gamma isomer)	0.004
Methaoxychlor (1,1,1-trichloro-2,2-bis[p-methoxy-phenyl] ethane)	0.1
Total Trihalomethanes	0.10
Toxaphene (C10H10CL8)-technical chlorinated camphene, 67-69 percent chlorine)	0.005
2,4-D (2,4-dichlorophenoxyacetic acid)	0.1
2,4,5-TP Silvex (2,4,5-trichlorophenoxypropionic acid)	0.01

(D) FLUORIDE QUANTITY

(1) Bottled water packaged in the United States, to which no fluoride is added, shall not contain fluoride in excess of the levels in Table 1.

TABLE 1:

Annual Average of Maximum Daily Air Temperatures (°F)	Fluoride Concentration in Milligrams Per Liter
53.7 and below	2.4
53.8 - 58.3	2.2
58.4 - 63.8	2.0
63.9 - 70.6	1.8
70.7 - 79.2	1.6
79.3 - 90.5	1.4

(2) Imported bottled water, to which no fluoride is added, shall not contain fluoride in excess of 1.4 mg/L.

(3) Bottled water packaged in the United States, to which fluoride is added, shall not contain fluoride in excess of levels in Table 2.

TABLE 2.

Annual Average of Maximum Daily Air Temperatures (°F)	Fluoride Concentration in Milligrams Per Liter
53.7 and below	1.7
53.8 - 58.3	1.5
58.4 - 63.8	1.3
63.9 - 70.6	1.2
70.7 - 79.2	1.0
79.3 - 90.5	0.8

(4) Imported bottled water, to which fluoride is added, shall not contain fluoride in excess of 0.8 mg/L.

(E) RADIOLOGICAL QUALITY

(1)

CONTAMINANT	PICOCURIES/L
Gross Alpha Acitivity	15
Radium - 226 & 228	5

(2) The bottled water shall not contain beta particle and photon radioactivity from manmade radionuclides in excess of that which would produce an annual dose equivalent to the total body or any internal organ of 4 millirems per year calculated on the basis of an intake of 2 liters of the water per day. If two or more beta or photon- emitting radionuclides are present, the sum of their annual dose equivalent to the total body or to any internal organ shall not exceed 4 millirems per year.

BIBLIOGRAPHY

ABC News. "Water - A Clear and Present Danger." ABC News. August 5, 1983. Show #119. Transcript 11p.

Anderson, A.C., Reimers, R.S., DeKernion, P. "A Brief Review of the Current Status of Alternatives to Chlorine Disinfection of Water." Am. J. Public Health. 72 (1982):1290-1293.

Anderson, T.W. "Water Hardness and Heart Disease." J. Am. Water Works Assoc. 68(July 1976):21.

Arehart-Treichel, J. "Eating Your Way Out of High Blood Pressure." Science News. 123(April 9, 1983):232-233.

Armstrong, B.K., McCall, M.G., Binns, C.W., Campbell, N.A., Masarei, J.R.L. "Water Sodium and Blood Pressure in Rural School Children." Archives of Environmental Health. 37(July/August 1982):236-245.

Ashmead, D. Chelated Mineral Nutrition. Huntington Beach, CA:International Institute of Natural Health Sciences, 1981.

Au, L.K.L. Personal Communication. January 7, 1983. Letter.

Banik, A.E. The Choice is Clear. Raytown, MI:Acres U.S.A., 1975.

Beaumen, W.H. "Purifiers." Everpure Bulletin. September, 1983.

Blackburn, H., Prineas, R. "Diet and Hypertension: Anthropology, Epidemiology and Public Health Implications." Prog. Biochemical Pharmacology. 19(1983):31-79.

Borgman, R.F., Lightsey, S.F. "Effects of Synthesized Hard Water and of Cadmium in the Drinking Water Upon Lipid Metabolism and Cholelithians in Rabbits." Am. J. of Veterinary Research. 43(August 1982):1432-1435.

Bragg, P.C., Bragg, P. The Shocking Truth About Water: The Universal Fluid of Death. Santa Barbara, CA:Health Science, 1977.

Bryant, G., Jordan, R. "Effect of Different Cooking Waters on Calcium Content in Certain Vegetables." Food Research. (1948):308-314.

Bukro, C. "Judge's Ruling Muddies Illinois Drinking Water." Health Freedom News. April 1983:28-31.

Burk, D. "Fluoridation: A Burning Controversy." Bestways. April 1982:40-44.

Burk, D. Personal Communication. May 1983. Letter.

Burton, A.C., Cornhill, F. "Correlation of Cancer Death Rates With Altitude and With the Quality of Water Supply of the 100 Largest Cities in the United States." J. Toxicology and Environmental Health. 3(1977):465-478.

Burton, A.C., Cornhill, F., Canham, P.B. "Protection from Cancer by 'Silica' in the Water-Supply of U.S. Cities." J. Environmental Pathology and Toxicology. 4(1980): 31-40.

Business Week. "Water's Rise: A Sales Torrent." Business Week. January 11, 1982:97-99.

Calabrese, E.J., Tuthill, R.W., Sieger, T.L., Klar, J.M. "The Role of Elevated Levels of Sodium in Diet and Drinking Water on the Development of Hypertension in Animals and Humans." J. of Environmental Pathology and Toxicology. 3(1980):143-150.

Carlo, G.L., Mettlin, C.J. "Cancer Incidence and Trihalomethane Concentrations in a Public Water System." Am. J. Public Health. 70(May L980):523-525.

Chambers, B. Personal Communication. November 3, 1983. Interview.

Changing Times. "Home Filters to 'Purify' Water." Changing Times. February 1981:44-46.

Churchill, D., Bryant, D., Fodor, G., Gault, M.H. "Drinking Water Hardness and Urolithiasis." Annals of Internal Medicine. 88(April 1978):513-514.

Clark, L. The Best of Linda Clark: Some Unusual Approaches to Health. New Canaan, CT:Keats, 1976.

Comstock, G.W. "Reviews and Commentary: Water Hardness and Cardiovascular Diseases." Am. J. Epidemiology. 110(October 1979):375-400.

Conforti, P.M., Kanarek, M.S., Jackson,L.A., Cooper, R.C., Murchio, J.C. "Asbestos in Drinking Water and Cancer in the San Francisco Bay Area: 1969-1974 Incidence." J. Chronic Diseases. 34(1981):211-224.

Consumer Reports. "The Selling of H20." Consumer Reports. September 1980:531-538.

Consumer Reports. "Water Filters." Consumer Reports. February 1983:68-73, 102.

Crawford, M.D., Gardner, M.J., Morris, J.N. "Cardiovascular Disease and the Mineral Content of Drinking Water." British Medical Bull. 27(1971):21-24.

Crawford, M.D., Gardner, M.J., Morris, J.N. "Changes in Water Hardness and Local Death-Rates." Lancet. August 1971:327-329.

Dauncey, M.J., Widdowson, E.M. "Urinary Excretion of Calcium, Magnesium, Sodium, and Potassium in Hard and Soft Water Areas." Lancet. April 1972:711-715.

Dawson, E.B., Fieve, R., Sheard, M.L. "Shall We Add Lithium to Drinking Water?" International Water Quality Symposium ed. by D.X. Manners:47-52.

Dawson, E.B., Frey, M.J., Moore, T.D., McGanity, W.J. "Relationship of Metal Metabolism to Vascular Disease Mortality Rates in Texas." Am. J. Clinical Nutrition. 31(JUly 1978):1188-1197.

Deal,W.F. Personal Communication. November 1, 1983. Phone Conversation.

Department of Consumer Affairs. Clean Your Room! Compendium on Indoor Pollution. Sacremento, CA:State of California, 1982.

Donsbach, K.W., Walker, M. Drinking Water. Huntington Beach, CA:International Institute of Natural Health Sciences, 1981.

Dougherty, E. You Have a Right to Know. Lincoln, NE:Kane Associates, 1980.

Durfor, C.N., Becker, E. Public Water Supplies of the 100 Largest Cities in the United States, 1962. Washington:GPO, 1964.

Edelman, Daniel J., Inc. "Information Packet on Evian Water." New York, 1982.

EPA. Bottled Water. Criteria and Standards Division. Office of Drinking Water. 1p. Not Dated.

EPA. Bottled Water Study: A Pilot Survey of Water Bottlers and Bottled Water. Washington:GPO, 1972.

EPA. "Drinking Water and Health: Recommendations of the National Academy of Sciences." Federal Register. 42(July 11, 1977):35764-35779.

EPA. A Drop to Drink: A Report on the Quality of Our Drinking Water. Washington: GPO, 1976.

EPA. Home Drinking Water Treatment Units Contract. Criteria and Standards Division, Office of Drinking Water. Phase 1 and 2:July 1980:Phase 3; March 1982.

EPA. Water Programs. "National Interim Primary Drinking Water Regulations." Federal Register. December 24, 1975; July 9, 1976; November 29, 1979; March 11, 1980, August 27, 1980; March 3, 1982; March 12, 1982. Sections 141.1-141.42.

Elinder, C., Stenstrom, T., Piscator, M., Linnman, L., Jonsson, L. "Water Hardness in Relation to Cadmium Accumulation and Microscopic Signs of Cardiovascular Disease in Horses." Archives of Environmental Health. 35(March/April 1980):81-84.

Epstein, S.S., Zavon, M. "Is There a Threshold for Cancer." International Water Quality Symposium ed. by D.X. Manners:54-62.

FDA. "Quality Standards for Foods with No Identity Standards." (Bottled Water). Federal Register. 42(March 15, 1977):14325 and 44(March 6, 1979):12172.

Faelten, S. The Complete Book of Minerals for Health. Emmaus, PA:Rodale, 1981.

Faust, H.S. "Effects of Drinking Water and Total Sodium Intake on Blood Pressure." Am. J. of Clinical Nutrition. 35(June 1982):1459-1467.

Fry, T.C., Shelton, H., Esser,W., Kloss, J., Tilden, J.H., Vance, J.A., Carque, O., Carrington, H., Bernard, R.W. The Great Water Controversy. Yorktown, TX: Life Science, N.D.

Furman, A.F. "Fluoridation Boon or Bane." Health Express. (October 1982):18+.

Gottlieb, M.S., Carr, J.K., Morris, D.T. "Cancer and Drinking Water in Louisiana: Colon and Rectum." International J. Epidemology. 10(June 1981):117-125.

Greathouse, D.G., Osborne, R.H. "Preliminary Report on Nationwide Study of Drinking Water and Cardiovascular Diseases." J. of Environmental Pathology and Toxicology. 3(1980):65-76.

Greenberg, L.A., Turner, J.M. "Effect of Carbonated Water on Gastric Acid Secretion." New England J. Medicine. 253(1955):105-106.

Hallenbeck, W.H., Brenniman, G.R., Anderson, R.J. "High Sodium in Drinking Water and Its Effect on Blood Pressure." Am. J. Epidemiology. 114(1981):817-825.

Hanway, J.J., Herrick, J.B., Waillrich, T.L., Bennett, P.C., McCall, J.T. The Nitrate Report. Ames, Iowa: Iowa State University, 1963.

Haring, B.J.A., Zoeteman, B.C.J. "Corrosiveness of Drinking Water and Cardiovascular Disease Mortality." Bull. Environmental Contaminant Toxicology. 25(1980):658-662.

Haring, B.S.A., Van Delft, W. "Changes in the Mineral Composition of Food as a Result of Cooking in 'Hard' and 'Soft' Waters." Archives of Environmental Health. 36(January/February 1981):33-35.

Health Freedom News. "Suspected Carcinogen Found in Water." Health Freedom News. (September 1982):16.

Heyden, S. "The Hard Facts Behind the Hard-Water Theory and Ischemic Heart Disease." J. Chronic Diseases. 29(1976):149-157.

Hoffman, A., Valkenburg, H.A., Valkenburg, G.J. "Increased Blood Pressure in Schoolchildren Related to High Sodium Levels in Drinking Water." J. of Epidemology and Community Health. 34(1980):179-181.

Hogan, M.D., Chi, P., Hoel, D.G. "Association Between Chloroform Levels in Finished Drinking Water Supplies and Various Site-Specific Cancer Mortality Rates." J. Environmental Pathology and Toxicology. 2(1979):873-887.

Ingols, R.S., Craft, T.F. "Analytical Notes-Hard- vs. Soft-Water Effects on the Transfer of Metallic Ions From Intestine." J. Am. Water Works Assoc. 68(April 1976): 209-210.

Jansson, B. "Seneca County, New York: An Area with Low Cancer Mortality Rates." Cancer. 48(1981):2542-2546.

Jensen, O.M. "Nitrate in Drinking Water and Cancer in Northern Jutland, Demark, with Special Reference to Stomach Cancer." Ecotoxicology and Environmental Safety. 6(1982):258-267.

Keough, C. Water Fit to Drink. Emmaus, PA:Rodale, 1980.

Lancet. "Cancer and Chlorinated Water." Lancet. May 23, 1981:1142.

Lolli, G., Greenberg, L.A., Lester, D. "The Influence of Carbonated Water on Gastric Emptying." New England J. Medicine. 246(1951):490-492.

Manners, David X. ed. International Water Quality Symposium: Water, Its Effects on Life Quality. Proceedings of the Seventh International Water Quality Symposium: Wash., D.C.:Water Quality Research Council, 1974.

Masironi, R., Hamilton, E.I., Lew, E., Armstrong, R. "The Geography of Death." International Water Quality Symposium ed. by D.X. Manners:171-173.

Masironi, R., Shaper, A.G. "Epidemiological Studies of Health Effects of Water from Different Sources." Annual Review of Nutrition. 1(1981):375-400.

Maugh, T.H. "New Study Links Chlorination and Cancer." Science. 211(February 13, 1983):694.

McCauley, P.T., Bull, R.J. "Experimental Approaches to Evaluating the Role of Environmental Factors in the Development of Cardiovascular Disease." J. Environmental Pathology and Toxicology. 4(September 1980):27-50.

Mertz, W. "Chromium, Selenium and Other Essential Trace Elements." International Water Quality Symposium ed. by D.X. Manners:41-43.

Morales, B.L., Clark, J.G. "Longevity, Plus-I, II, III." Organic Consumer Report. 62 (September 14, 21 and 27, 1982)#37, 38 and 39.

Morell, F. The Bio-Electronic Method of Prof. Vincent. Transcript of Lecture. Occidental Institute Alumni Assoc. San Bruno, CA:Occidental Institute Alumni Assoc., 1982.

Muehling, E.C. Water for the Eighties: A Cause for Concern. Lincoln, NE:Kane Assoc., 1976.

Nacewicz, R. Personal Communication. November 1, 1983. Phone Conversation.

National Research Council. Drinking Water and Health. V.1(1977), V.2(1980), V.3(1980), V.4(1982) Wash., D.C.:National Academic Press.

Neal, J.B., Neal, M. "Effect of Hard Water and MgSO4 on Rabbit Atherosclerosis." Archives of Pathology. 73(May 1962):58-61.

Neri, L.C., Hewitt, D., Schreiber, G.B., Mandel, J.S. "Is There a Water Factor: A Case for Magnesium." International Water Quality Symposium. ed. by D.X. Manners:73-75.

Neri, L.C., Hewitt, D., Schreiber, G.B. "Can Epidemiology Elucidate the Water Story." Am. J. Epidemiology. 99(February 1974):75-88.

Neri, L.C., Hewitt, D., Schreiber, G.B., Anderson, T.W., Mandel, J.S., Zdrojewsky, A. "Health Aspects of Hard and Soft Waters." J. Am. Water Works Assoc. 67(August 1975):403-409.

Ohanian, E.V., Cirolia, D.M. "Sodium in Drinking Water as a Etiological Factor in Hypertension." Water Technology. (October 1983):28-36.

Page, T., Harris, R.H., Epstein, S.S. "Drinking Water and Cancer Mortality in Louisiana Science. 193(July 1976):55-57.

Passwater, R. Supernutrition for Healthy Hearts. New York:Jove, 1978.

Perry, H.M., Perry, E.F., Erlanger, M.W. "Possible Influence of Heavy Metals in Cardio vascular Disease: Introduction and Overview." J. Environmental Pathology and Tox icology. 3(1980):195-203.

Pocock, S.J., Shaper, A.G., Cook, D.G., Packham, R.F., Lacey, R.F., Powell, P., Russell, P.F. "British Regional Heart Study: Geographic Variations in Cardiovascular Mortality and the Role of Water Quality." British Medical J. 24(May 1980): 1243-1249.

Prevention. "Your Government and Your Health." Prevention. September 1983:169.

Price, J.M. Coronaries/Cholesterol/Chlorine. New York:Pyramid, 1969.

Price, W.A. "Race Decline and Race Regeneration." J. Am. Dental Assoc. 28(April 1941):548-558.

Price, W.A. Nutrition and Physical Degeneration. San Diego: Price-Pottenger, 1977.

Puddu, V., Menotti, A., Signoretti, P. "Drinking Water and Cardiovascular Disease." Am. Heart J. 99(April 1980):539-540.

Rand, J. Water. Unpublished paper. 43p. 1982.

Robertson, J.S., Slattery, J.A., Parker, V. "Water Sodium, Hypertension and Mortality." Community Medicine. 1(1979):295-300.

Sauer, H.I. "Relationship of Water to Risk of Dying." International Water Quality Symposium ed. by D.X. Manners:76-79.

Schroeder, H.A. "Relation Between Mortality from Cardiovascular Disease and Treated Water Supplies." J. Am. Medical Assoc. (April 23, 1960:98-104.

Schroeder, H.A. "The Role of Trace Elements in Cardiovascular Diseases." Medical Clinics of North America. 58(March 1974):381-396.

Schroeder, H.A. The Poisons Around Us. New Canaan, CT:Keats, 1974.

Schwartz, S. The Book of Waters. New York:A & W Publishers, 1979.

Science News. "Calcium, Chlorine and Heart Disease Linked." Science News. (August 13, 1983):103.

Scott, F.I. "Editor's Page." Review of Fluoridation on Tap by G. Walker. Am. Laboratory. 15(April 1983):6-8.

Selikoff, I.J. "Asbestos in Water." International Water Quality Symposium ed. by D.X. Manners:66-70.

Shaper, A.G., Pocock, S.J., Walker, M., Cohen, N.M., Wade, C.J., Thomson, A.G. "British Regional Heart Study: Cardiovascular Risk Factors in Middle-aged Men in 24 Towns." British Medical J. 283(July 1981):179-186.

Sharrett, A.R. "The Role of Chemical Constituents of Drinking Water in Cardiovascular Diseases." Am. J. Epidemiology. 110(October 1979):401-419.

Sharrett, A.R., Heyden, S., Masironi, R., Greathouse, D., Shaper, A., Hewitt, D. "Panel Discussion: The Relationship of Hard Water and Soft Water in CVD and Health." J. Environmental Pathology and Toxicology. 4(1980):113-141.

Sierakowski, R., Finlayson, B., Landes,R. "Stone Incidence as Related to Water Hardness in Different Geographical Regions of the United States." Urological Research. 7 (1979):157-160.

Sigurdson, E.E., Levy, B.S., Mandel, J., McHugh, R., Michienzi, L.J., Jagger, H., Pearson, J. "Cancer Morbidity Investigations: Lessons from the Duluth Study of Possible Effects of Asbestos in Drinking Water." Environmental Research. 25 (1981):50-61.

Sorenson, J. Personal Communication. November 3, 1983. Phone Conversation.

Studlick, J.R.L., Bain, R.C. "Bottled Waters - Expensive Ground Water." Ground Water. 18(July/August 1980): 340-345.

Studlick, J., Bain, R. "Bottled Water; Expensive Ground Water." Water Well J. (July 1980):75-79.

Suffolk County Department of Health Services. 1982 Report on Bottled Water and Bottled Water Substitutes. Hauppauge, New York:Suffolk County, 1982.

Thouez, J. Beauchamp, Y., Simard, A. "Cancer and the Physicochemical Quality of Drinking Water in Quebec." Social Science and Medicine. 15D(1981):213-223.

Tuthill, R.W., Calabrese, E.J. "Age as a Function in the Development of Sodium - Related Hypertension." Environmental Health Perspectives. 29(April 1979):35-43.

Tuthill, R.W., Calabrese, E.J. "Elevated Sodium Levels in the Public Drinking Water as a Contributor to Elevated Blood Pressure Levels in the Community." Archives of Environmental Health. 34(July/August 1979):197-203.

U.S. Code of Federal Regulations. Title 21:Part 103. Subpart B. Standards of Quality. Bottled Water (Paragraph 103.35):52-54. Part 129. Processing and Bottling of Bottled Drinking Water. 132-136. WAshington:GPO, 1983.

U.S. General Accounting Office. Report to the Administrator Environmental Protection Agency. States' Compliance Lacking in Meeting Safe Drinking Water Regulations. March 3, 1982. Washington:GPO, 1982.

Von Wiesenberger, A. Oasis: The Complete Guide to Bottled Water Throughout the World. Santa Barbara, CA:Capra Press, 1978.

Voors, A.W. "Minerals in the Municipal Water and Atherosclerotic Heart Death." Am. J. Epidemiology. 93(1971):259-266.

Waldbott, G.L., Burgstahler, A.W., McKinney, H.L. Fluoridation: The Great Dilemma. Lawrence, KS:Coronado Press, 1978.

Walker, Norman. Water Can Undermine Your Health. Phoenix, AZ:Woodside, 1974.

Wasserman, A.E. "Nitrates, Nitrites, and Nitrasmines." International Water Quality Symposium ed. by D.X. Manners:63-65.

Wilkins, J.R., Reiches, N.A., Kruse, C.W. "Organic Chemical Contaminants in Drinking Water and Cancer." Am. J. Epidemology. 110(October 1979):420-448.

Wilkins, J.R., Comstock, G.W. "Source of Drinking Water at Home and Site-Specific Cancer Incidence in Washington County, Maryland." Am J. Epidemiology. 114 (1981):178-190.

Willett, W.C. "Drinking Water Sodium and Blood Pressure: A Cautious View of the 'Second Look'." Am. J. Public Health. 71(July 1981):729-732.

Williams, R.T. "Implied Claims Verification Testing - Water Filters/Purifiers - 800530." General Ecology, Inc. 1980.

Williamson, S.J. "Epidemiological Studies on Cancer and Organic Compounds in U.S. Drinking Waters." Science of the Total Environment. 18(981):187-203.

Winton, E.F., McAbe, L.J. "Studies Relating to Water Mineralization and Health." J. Am. Water Works Assoc. 62(January 1970):26-30.

Yiamouyiannis, J.A. Everything You Wanted to Know About Fluoridation - But Were Afraid to Ask: A Discovery Deposition. Monrovia, CA: National Health Federation, 1977.

Yiamouyiannis, J.A. Fluoride: The Aging Factor. Delaware, Ohio: Health Action Press, 1983.

Yiamouyiannis, J.A. Lifesavers Guide to Fluoridation: Risks/Benefits Evaluated in this 1982 Question and Answer Report. Delaware, Ohio: Safe Water Foundation, 1982.

• BIOGRAPHY

Martin Fox, Ph.D., nutritionist, environmental researcher and author is currently living in Austin, Texas with his wife and son. He has been associated with the California Clinic of Preventative Medicine and The Nevada Clinic, both homeopathic and nutrition oriented medical clinics. He received his educational degrees from Louisiana State University and Donsbach University.